FRAMINGHAM STATE COLLEGE
3 3014 00055 1426

D1237833

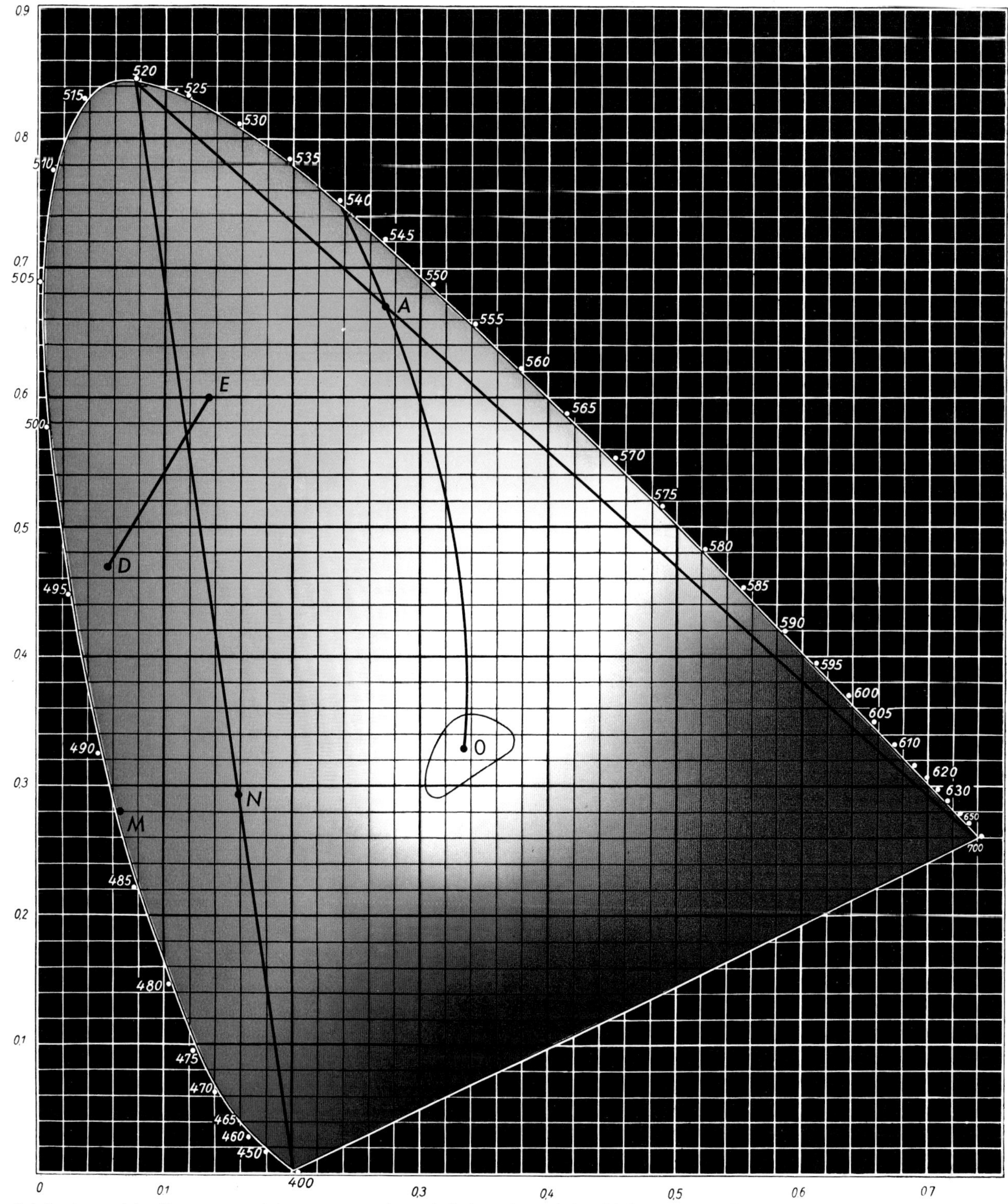

C. I. E. chromaticity diagram, containing representations of all the colors of the visible spectrum. For further explanation, see page 187.

A Course for Generalists

HYSICS FOR SOCIETY

W. B. PHILLIPS

University of West Florida

Pensacola, Florida

ADDISON-WESLEY PUBLISHING COMPANY, INC.

Reading, Massachusetts · Menlo Park, California

London · Don Mills, Ontario

This book is in the Addison-Wesley series in Physics

Consulting editor: David Lazarus

Library of Congress Catalog Card No. 70–140836.

PREFACE

The student for whom this book is intended has had little background in the sciences—probably none in physics—and he would not normally take a college physics course. He almost certainly would not take the usual introductory physics course, for this type of course fails to attract and interest him for many of the same reasons that he isn't a physics major in the first place.

This text is designed to meet the nonscience student halfway. It is a book *about* physics, from which the reader may learn some physics, rather than a traditional text *in* physics. Thus it can provide the student who has rejected the usual approach to physics—who dislikes mathematics and is bored to tears by levers and pulleys and inclined planes—the opportunity to learn something about physics and what it means.

This book makes no attempt to survey all of physics. Rather, those parts which are most interesting, important, and useful have been chosen to the exclusion of other, traditional topics. Much of mechanics, electrostatics, and geometrical optics is omitted, though such topics as thermoelectricity, lasers and cryogenics—which are not often included in introductory courses—are discussed. An attempt has been made to avoid telling the student more about a subject than he really wants to know, or is likely to find it useful to know.

There are no problems to be solved by mathematical manipulation. This does not mean that there are no problems to be solved, but that the problems are designed to challenge the student with ideas instead of equations. The problems chosen are drawn from the areas of the economic, sociological, and humanitarian—as well as technical—aspects of physics.

The illustrations and examples presented have been chosen from topics having to do with contemporary civilization—such things as H-bombs and radiation effects; ICBM's, MIRV's, and ABM's; space exploration and its costs; and the uses and effects of nuclear power. In the past, teachers of physics have often avoided discussing the applications of physics, to say nothing of getting students to ask questions about the implications of physical developments for our civilization. Here the instructor is presented with a number of logical points of departure from which he may discuss the physical aspects of current national problems, thus making the course as "relevant" as he desires.

A list of collateral readings is given at the end of each chapter. Some of the references include more detailed discussions of the topics presented, while other readings furnish introductions to allied areas of interest that were not discussed. The *questions* which follow each chapter are designed primarily to test the student's understanding of the material presented in that chapter. The *problems* should be considered as exercises that call for solutions just as if they required numerical answers. In solving these problems, the student may wish to apply the ideas, tools, and techniques he has acquired elsewhere during his university education, in the true spirit of the generalist.

Pensacola, Florida W. B. P.
January 1971

CONTENTS

CHAPTER 1 THE GRAVITATIONAL FORCE

1.1 GRAVITY

The most commonly experienced, yet weakest, of all the forces in nature is that of gravity. Not only do we constantly experience the effects of this force pulling our bodies toward the surface of the earth, but we make a measurement of its strength each time we step on a scale (Fig. 1.1). Some of the things we know about the gravitational force are summarized below.

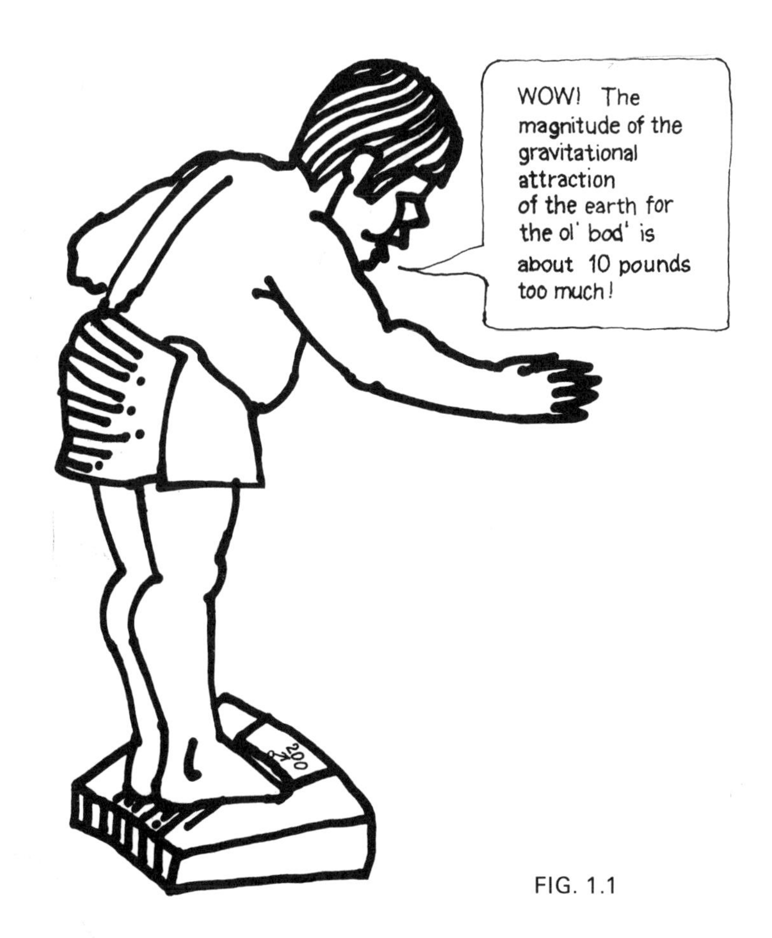

FIG. 1.1

The gravitational force is always attractive. Electrical forces may be either attractive or repulsive (unlike charges attract each other, like charges repel each other), but the gravitational interaction between two objects is always attractive. Thus we may first say that every bit of matter in the universe attracts every other bit of matter.

The gravitational force between any two objects depends upon the product of their masses. The size of the attractive force between any two objects is also determined by the quantities of matter, or masses, of those bodies. In comparison with the force acting between the earth and any object on it, the attractive force between two objects on the earth is

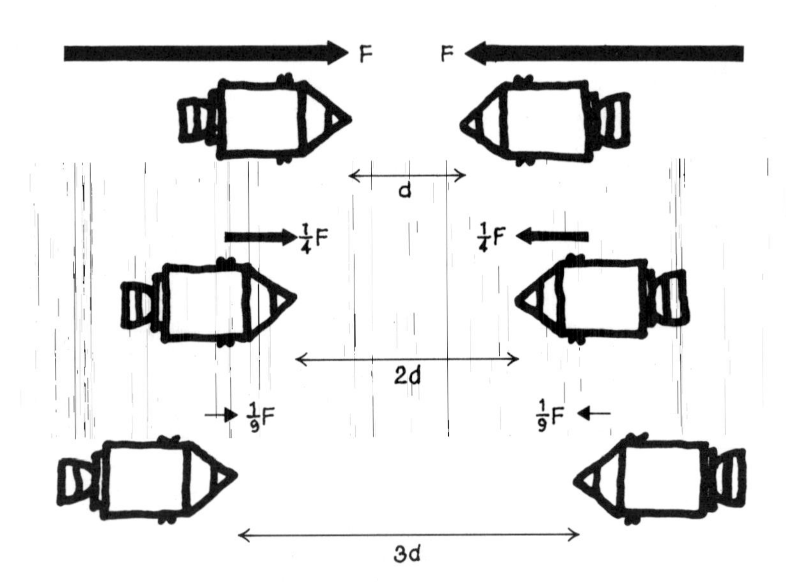

FIG. 1.2

extremely small, and can be measured only with great difficulty. Because the moon is less massive than the earth, astronauts who land there experience an attractive gravitational force only about one-sixth as great as they experience on earth; thus a 180-pound man (earth weight) weighs only about 30 pounds on the moon. The same man would weigh 475 pounds on the surface of the giant planet Jupiter.

The gravitational force obeys an inverse-square-law distance relationship. The attractive force between two objects diminishes in the same proportion as the square of the distance between them increases. This means that if the distance between two objects is doubled, the force between them will be only one-fourth as great as it was before; if the distance is tripled, the force will be only one-ninth as great (Fig. 1.2). Thus the gravitational attraction between objects decreases rapidly as the distance between them is increased.

The gravitational force is a long-ranged force. The attractive force between two objects does not fall to zero until the distance between them has become infinitely great. Incorrect descriptions of earth–moon voyages sometimes say that beyond a certain point the space vehicle has escaped the earth's gravity and is falling toward the moon. A vehicle cannot "escape" from the earth's gravity until it is infinitely distant from the earth. (What happens as that spaceship nears the moon is that the attraction of the moon's gravitational field for the ship becomes greater than that of the now-distant earth, but the earth continues to exert a force on the vehicle.)

We may finally put all these ideas together to say that every bit of matter in the universe attracts every other bit of matter with a force that is proportional to the product of their masses and inversely proportional to the square of the distance between them. This statement is known as the Law of Universal Gravitation.

1.2 THE ACCELERATION OF GRAVITY

So long as one is near the earth, one is attracted to it by the gravitational interaction between one's body and the earth. A person experiences this interaction as a *force* that is exerted on whatever object supports him. When he stands on that bathroom scale, the dial indicates the force with which he is being attracted by the earth. It is equally true that the

FIG. 1.3

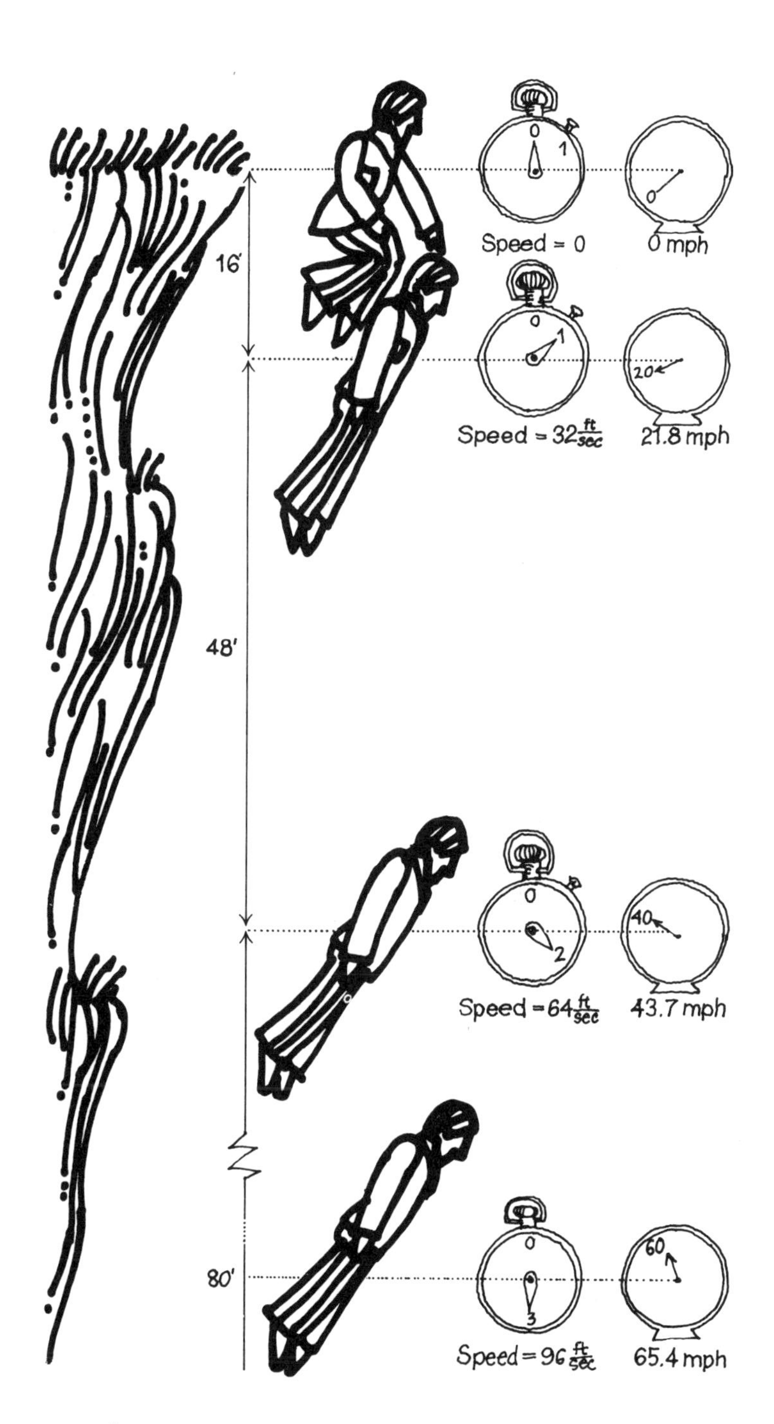

earth is attracted toward the person with the same force.

If a person near the earth were not supported by whatever he happened to be standing on, he would fall toward the earth with ever-increasing speed—he would be accelerated toward the earth's surface. This *acceleration due to gravity* is about 32 feet per second per second at sea level. This means that during each second you fall freely toward the earth your speed increases by 32 feet per second (almost 22 miles per hour). To illustrate what an acceleration of this magnitude means, suppose that you decided to "end it all" and threw yourself off the precipice at Lovers' Leap (Fig. 1.3).

If you jumped straight outward at the start of your fall, your downward speed would be zero, but, since you are being accelerated downward at 32 feet per second per second, at the end of the first second of your fall your speed would be 32 feet per second, about 21.8 miles per hour. During this first second of fall your *average* speed would have been 16 feet per second, so you would fall a distance of 16 feet during the first second. During the second second of your fall, your speed would increase by another 32 feet per second. At the end of two seconds you would be falling at 64 feet per second, or 43.7 miles per hour. Your average speed during this second second of fall would have been 48 feet per second; you would have fallen 48 feet, which is three times as far as you fell during the first second.

You may continue these calculations as long as you wish. At the end of the third second your speed would be 96 feet per second, 65.4 miles per hour. The point which should be apparent by now is that the acceleration due to gravity is a sizable effect. If your automobile were able to accelerate at a constant 32 feet per second per second, from a standing start you could reach 60 miles per hour in about 2.7 seconds and would be able to cover a quarter of a mile in slightly more than nine seconds.

Note that in all these calculations we have neglected the effect of air resistance on the falling body. Because of this air resistance, if you were to fall (or jump) from an airplane at a great altitude, your speed would not keep on increasing indefinitely. You would, instead, reach a *terminal velocity* of about 120 miles per hour. Fortunately, most other objects that fall from the sky also reach terminal velocities. If it were not slowed by air resistance, a raindrop formed at an altitude of 10,000 feet would be traveling at more than 550 miles per hour when it reached the earth's surface.

1.3 PROJECTILE MOTION

One reason all these facts about the acceleration due to gravity are important is that gravity is the determining factor in all types of projectile motion. Probably you were introduced to the practical problem of the *range* of a projectile when you were quite young and your concern was how far you could throw a baseball, or how far your friend could throw a snowball.

FIG. 1.4

You quickly discovered, by a process of experimentation, that if you threw the missile at too low an angle, it would not go very far before it hit the ground. Similarly, if the angle of projection became too great, the object would rise higher into the air but still would not go very far forward. Between these extremes there was a "just right" angle. This angle at which you could secure maximum range can be shown to be exactly 45 degrees (Fig. 1.4). Here, as in all our following discussion, the effects of air friction are neglected.

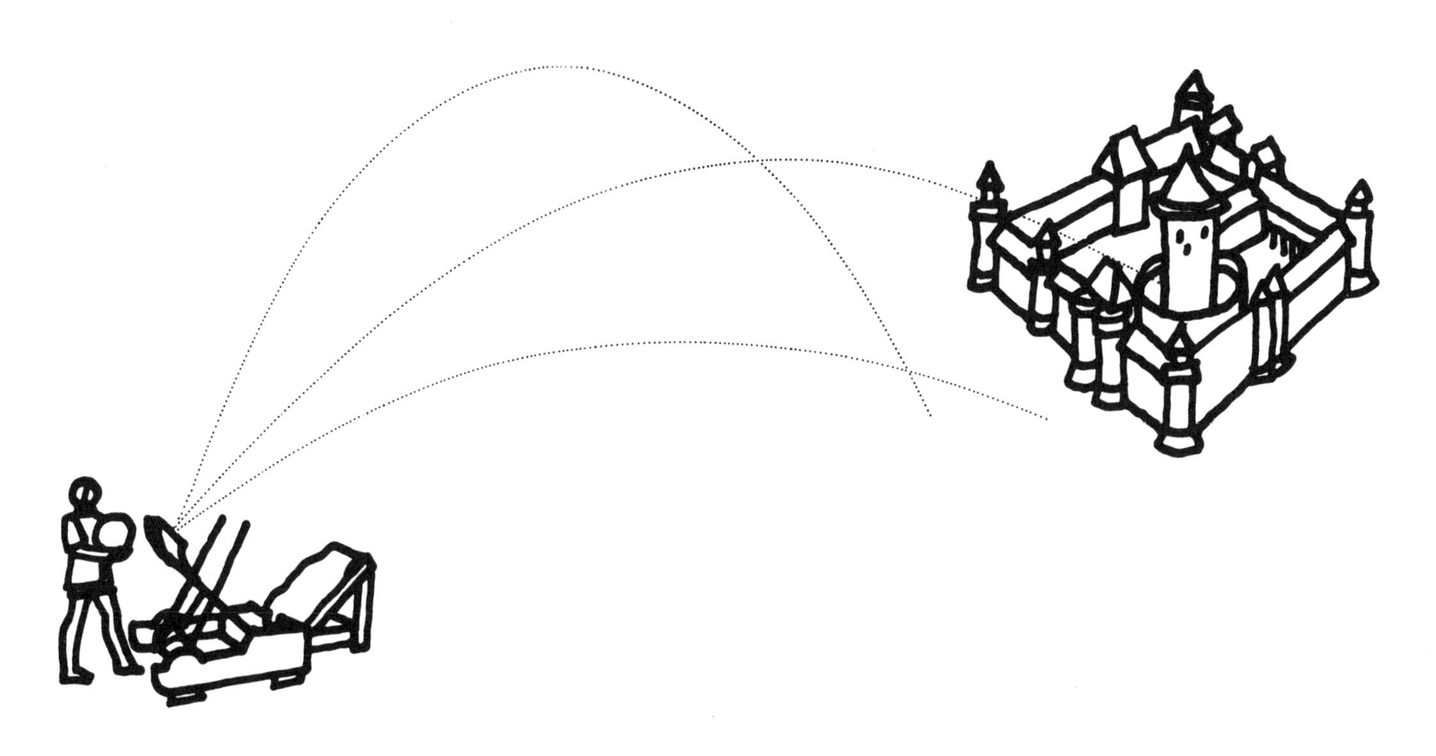

All projectiles near the earth's surface travel in *parabolic arcs.* A parabola is one member of an important family of curves we call *conic sections.* This name arises from the fact that if you were to take a slice exactly parallel to the side of a cone, the curve you would see left in the sliced cone would be a parabola (Fig. 1.5). Note the symmetry of this curve. The motion of any projectile may be considered in two symmetrical parts: going up and coming down.

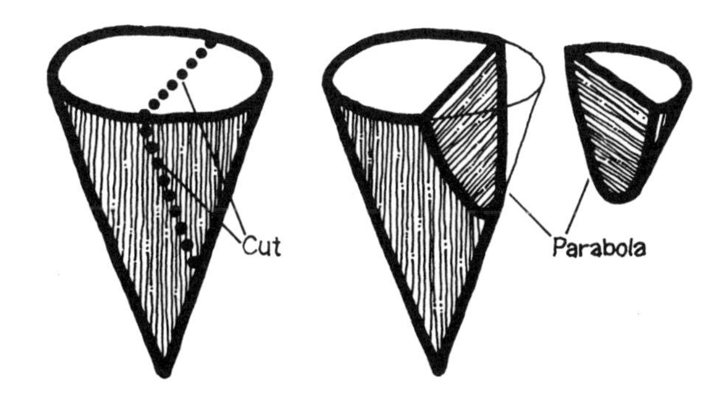

FIG. 1.5

We can further simplify the path of a projectile by separating its motion into *horizontal* and *vertical* components. To help you visualize

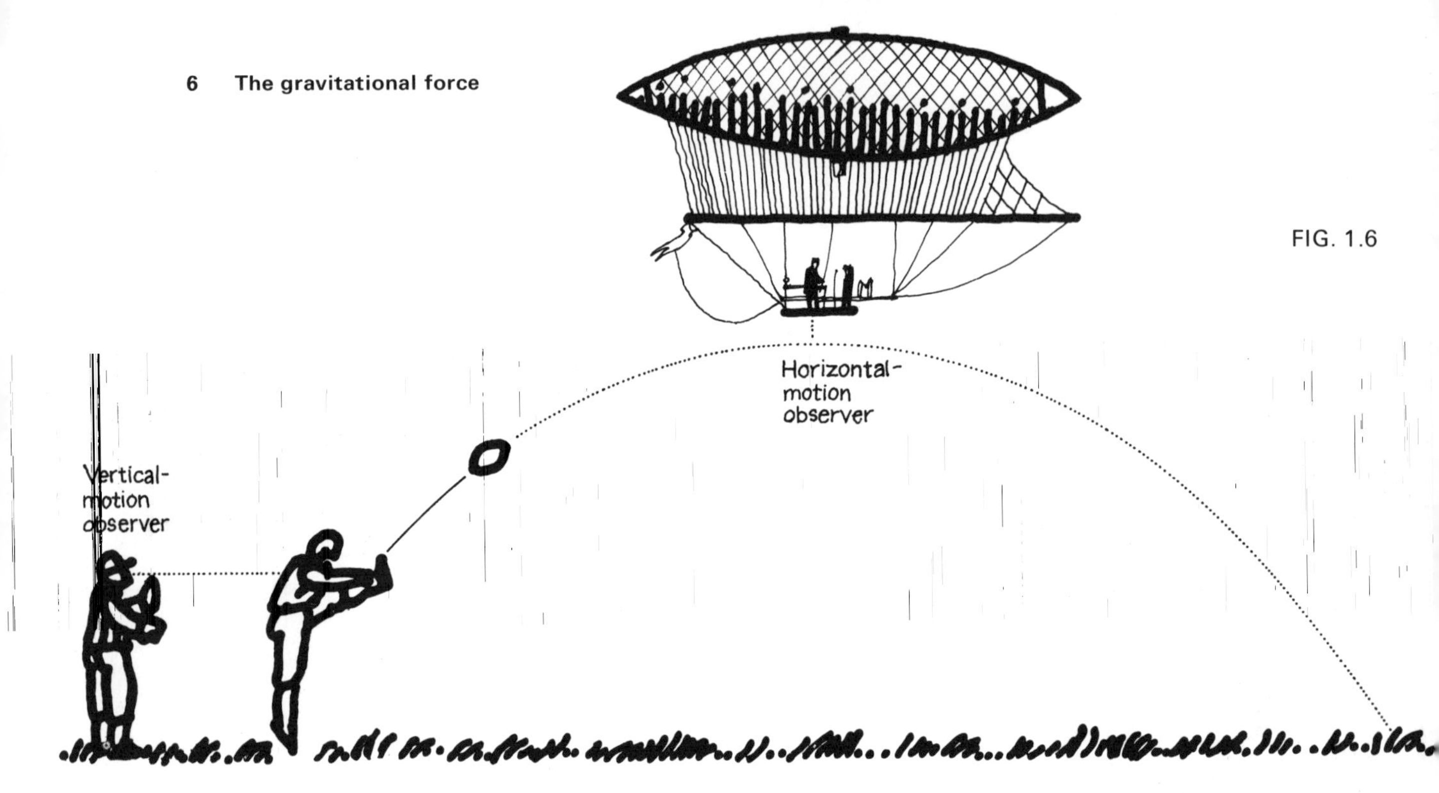

FIG. 1.6

this, suppose that you first stand directly behind someone who throws an object. Since you cannot readily detect the forward motion of the object from this position, you may see the object only go "straight up" and come back "straight down." You would have observed only the vertical component of the motion (Fig. 1.6).

Similarly, if you were above the path of the projectile and could not determine its changing height, you would see it move only with a constant horizontal speed. The horizontal motion is at a constant speed because the gravitational force is capable of modifying only the vertical part of the motion.

We may formalize the preceding statements by simply saying that the motion of any projectile is one of:

vertical motion with *constant acceleration,*
and
horizontal motion with *constant velocity.*

Now we may better understand why there should be an optimum angle for throwing a projectile for maximum range. If it is thrown at too low an angle, the object simply doesn't stay in the air long enough to go very far forward. If thrown at too great an angle, the object is in the air a long time but isn't going forward very fast during this time; hence it doesn't go very far forward.

1.4 ICBM'S, ABM'S, MIRV'S, AND ALL THAT

When launched at the proper angle and given enough speed, objects may be projected for very large distances in a gravitational field. In fact, it is quite possible to project a missile from any given spot on our planet and have it land at any other given spot on the globe. Missiles designed to do this while carrying

FIG. 1.7(a)

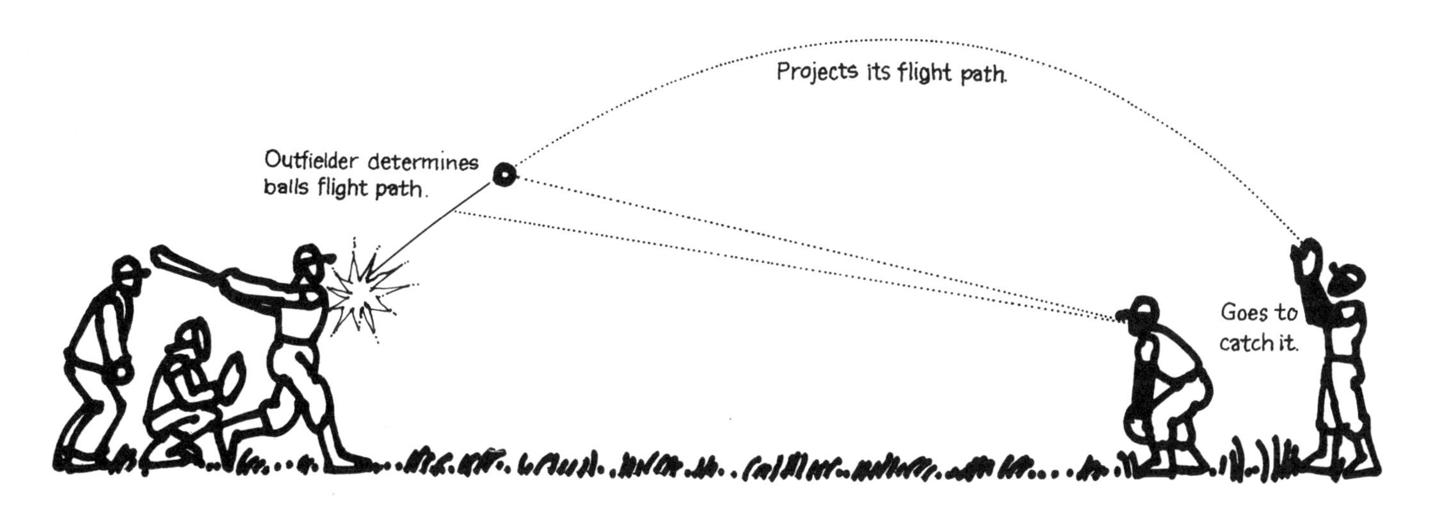

along a thermonuclear explosive device are usually called Intercontinental Ballistic Missiles (ICBM's). The motion of these missiles is simply projectile motion which is determined by the gravitational interaction. Thus the path of such a missile may be readily determined after it has been tracked by radar, or other means, for a short distance. The launching of a defensive missile designed to intercept and destroy the attacker is the objective of an Anti-Ballistic Missile (ABM) system.

The technique that may be employed to intercept an incoming ICBM is similar in many respects to that used by an outfielder to catch a fly ball. At the crack of the bat, the outfielder must quickly make an estimate of the ball's speed and direction, then determine its flight path and start toward the spot where he expects the catch to be made. From experience he knows that the ball will follow a modified (by air resistance) parabolic arc and, by tracking the ball during a small part of this arc, the outfielder mentally constructs the remainder of the ball's path. Of course the outfielder continues to visually track the ball during its flight and he uses this information to constantly update his estimate of its landing point and make appropriate corrections in the direction he is running (Fig. 1.7a).

An ICBM may be detected and tracked by radar relatively early in its trajectory. Its flight path may be calculated and—if all goes well—a very fast interceptor missile will be launched to rendezvous with the incoming missile and destroy it (Fig. 1.7b). Since an ICBM would be moving very rapidly as it approached its intended target, all detection and interception systems have to be computer controlled, for no human brain can respond quickly enough to direct these operations. The ABM system defensive missiles may themselves carry nuclear weapons; thus the result of the most

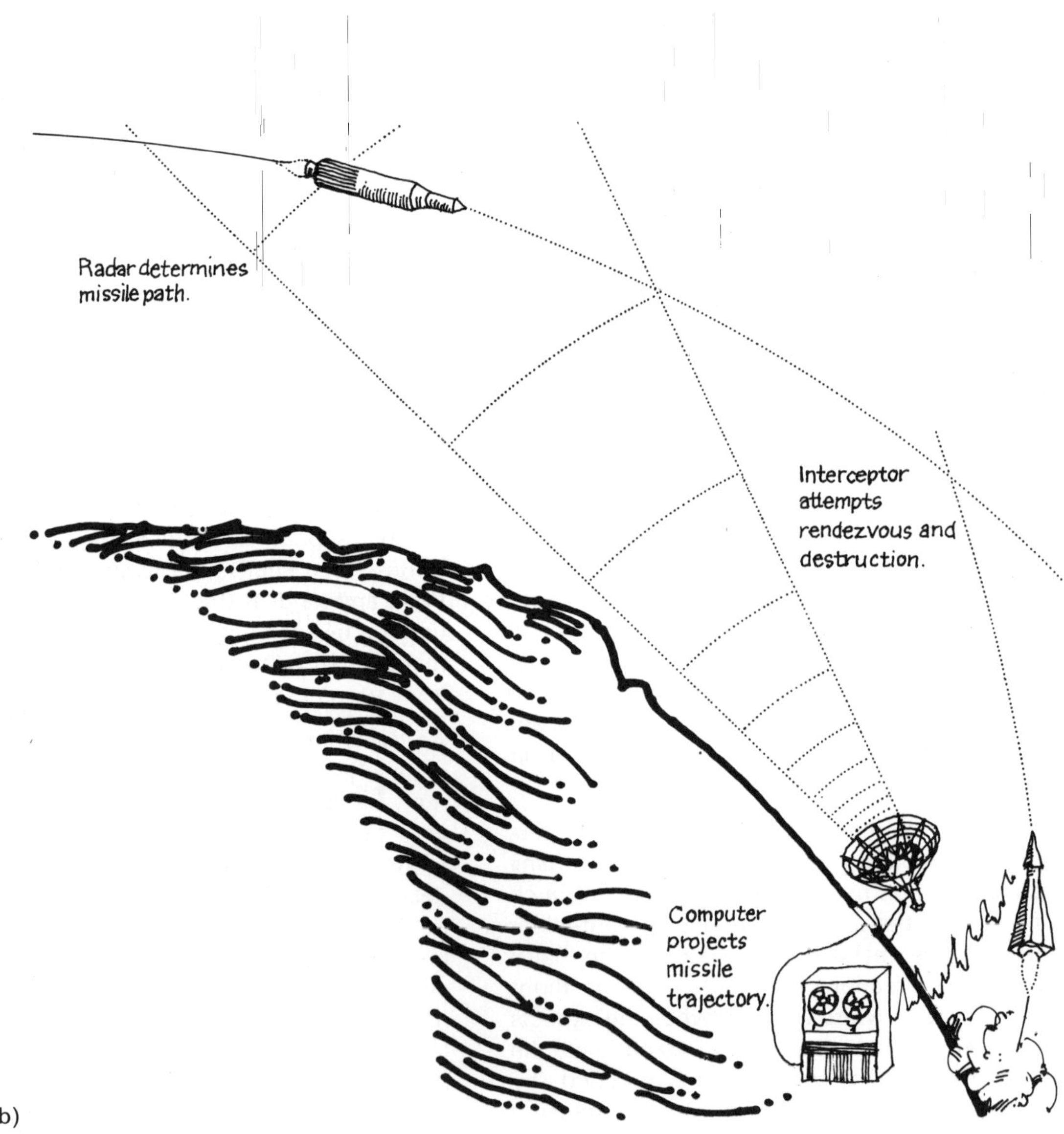
Radar determines missile path.
Interceptor attempts rendezvous and destruction.
Computer projects missile trajectory.

FIG. 1.7(b)

successful defensive system would be the detonation of H-bombs high above an intended target rather than at the target itself.

Naturally, a potential attacker would seek to thwart any ABM system. One means of doing this would be to equip each outgoing ICBM with multiple warheads. These Multiple Reentry Vehicle (MRV) warheads would then each have to be destroyed, thus greatly increasing the problems of the defending system. A still more sophisticated approach would be to employ a number of warheads, each carrying some sort of steering device, so that they could independently head for different targets. This Multiple Independently Targetable Reentry Vehicle (MIRV) system would force the defenders to deal with each of the new trajectories of the MIRV vehicles—the attacker would hope—so near the target that not enough time would be left for interception (Fig. 1.8). The game would be played in a still more complicated manner if the attacking missile were first put into a partial earth orbit, from which it could be called down upon its target. This Fractional Orbit Bombardment System (FOBS) might be made to approach the nation under attack from almost any direction with minimal advance warning. Of course a MIRV system could be incorporated into the FOBS.

The possible effectiveness of any antimissile system is still a matter for debate in the United States, although a limited ABM system has been deployed. Opponents argue that no ABM system will ever provide real protection, for the "offense" will always be able to keep a step ahead of the "defense." It is further contended that there is no way to test the

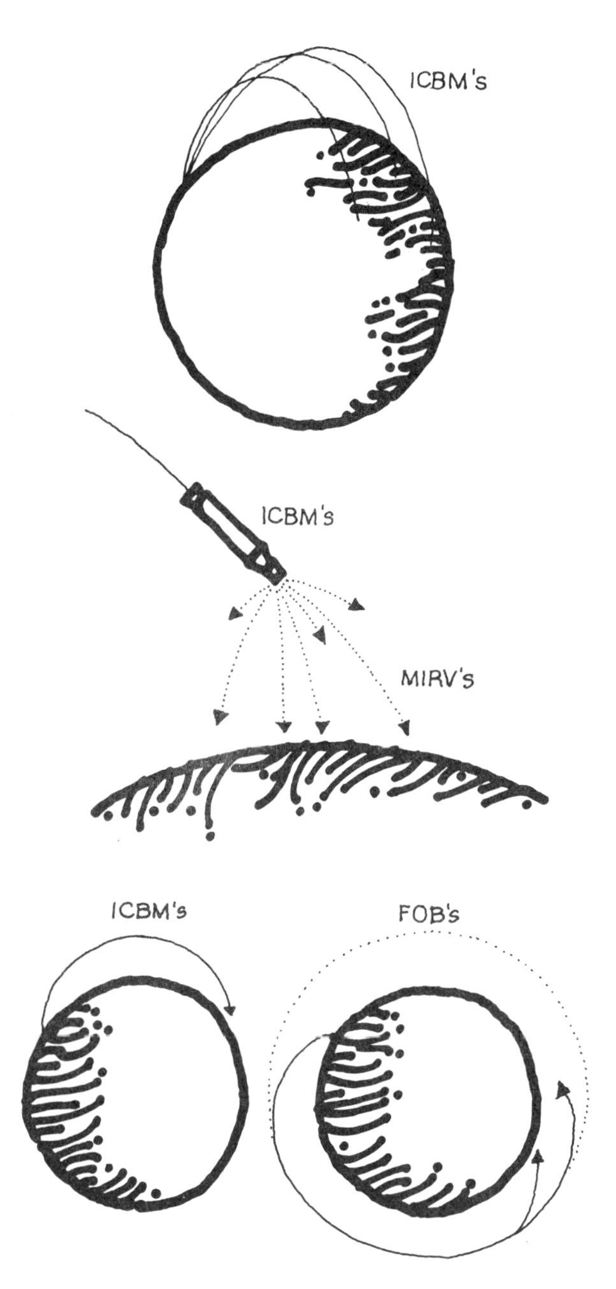

FIG. 1.8

actual workability of the system except under enemy attack, and grave doubts exist about the ability of a radar guidance system to continue to operate properly once H-bomb explosions begin to ionize the atmosphere. ABM proponents, however, deny these charges and reason that lack of any protective system might leave the nation subject to nuclear blackmail.

1.5 SATELLITES

As long ago as the seventeenth century, Sir Isaac Newton suggested that if one were to fire a projectile horizontally from the top of a high mountain at a great enough speed, it would circle the earth without falling to the ground (Fig. 1.9). In other words, it would go into orbit about the earth. Actually an orbit as close to the surface of the earth as even the height of the highest mountain could not be a stable one, since there is so much atmospheric drag at that height that the missile would slow and fall back to the ground. However, at altitudes above about 75 miles, the atmosphere is so thin that objects may be put into stable orbits about the earth.

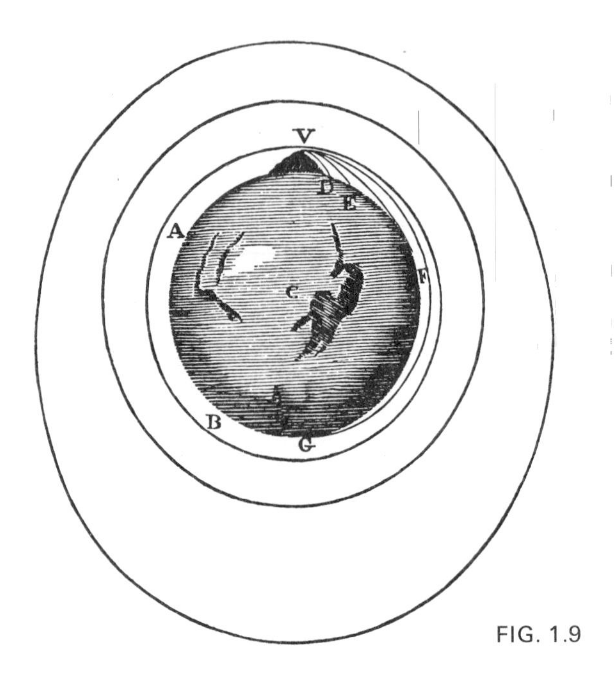

FIG. 1.9

Any object in an earth orbit, like our natural satellite the moon, is continually falling toward the earth, but if it has the proper speed, it can continue to move in its orbit and not come any closer to the earth. To see how this is possible, consider Fig. 1.10, which gives an approximate representation of this situation.

The object shown is assumed to be moving in a stable circular orbit about the earth. At any given instant it can be moving in only a single direction. Thus, if the object were to continue to move in that direction, it would necessarily move in a straight line such as A. But our satellite is continually falling toward the earth, and if we observe this object during a very short time interval we will see it fall a small distance B toward the earth. Since motion along line A from position 1 would have carried the object farther from the earth, at position 2 the object is at the same distance from the earth as at position 1 only because it fell back toward the earth.

At position 2 the object is again moving in a straight line (C), which would also carry it farther from earth. However, in the next short time interval, it will fall the same short distance B toward the earth, arriving at position 3

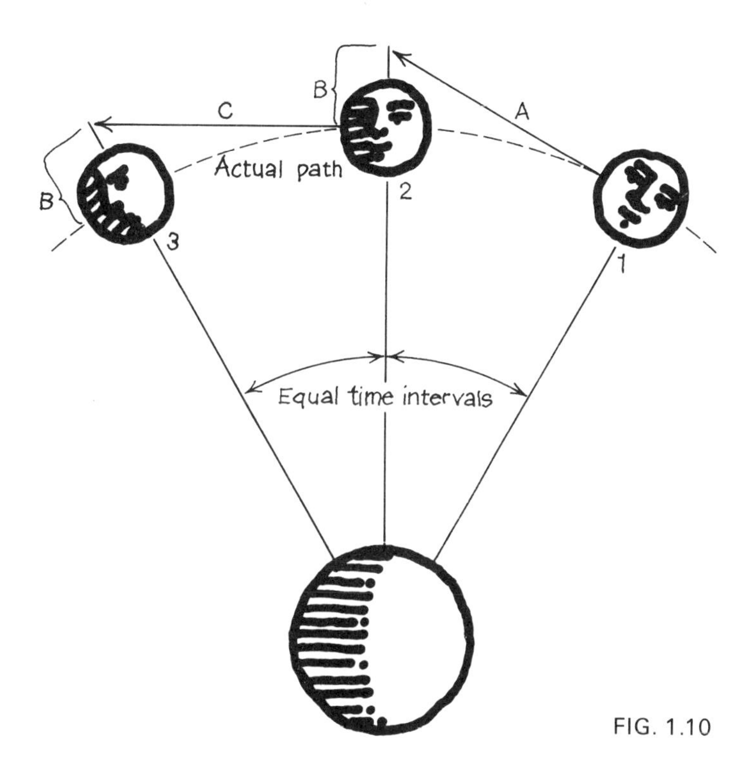

FIG. 1.10

still at the same height above earth as at positions 1 and 2.

You must not get the idea that the actual motion of the satellite is discontinuous or jerky, for we have divided this motion into segments only for the purpose of our illustration. If you wish, imagine that we use successively smaller and smaller time intervals in our observations. This would mean that the segments of the satellite path would become shorter and shorter. If this process were continued until the time intervals approached zero, the path of the motion would become a smooth, continuous curve.

Since an astronaut in earth orbit is continually falling toward the earth, he experiences zero gravity, or weightlessness, in the same way a person on earth would if he were inside a freely falling elevator. Because every object inside a space vehicle is also falling toward the earth, objects are not held in place by gravitational attraction and tend to float about. Inside a space vehicle there is no "up" or "down," except that determined by visual references.

We have learned that the gravitational attraction between any two objects decreases as the distance between them increases. In the case

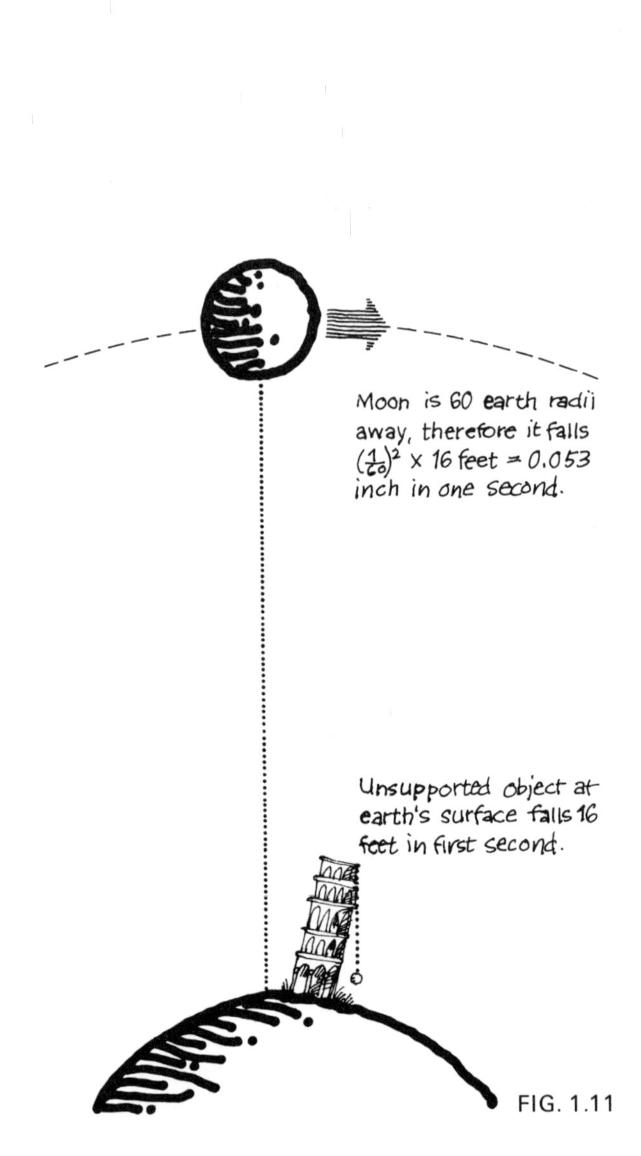

FIG. 1.11

of a satellite and the earth, the higher the satellite above the earth, the shorter the distance it would fall toward the earth during each time increment of its orbital motion. At the earth's surface an object falls 16 feet in the first second of free fall, while the moon, which is located about 60 earth radii distant, falls 1/3600 as far in one second, about 0.053 inch (Fig. 1.11). (Remember, the gravitational attraction depends on the inverse of the *square* of the distance.) Therefore the higher the satellite above the earth, the slower it needs to travel to stay exactly in orbit. This fact leads to the paradoxical situation that a satellite that has been moved from a lower orbit to a higher one will have had its orbital speed decreased, while a satellite that moves into a lower orbit will have had its speed increased. A satellite in orbit near the earth's surface must travel at more than 17,000 miles per hour to stay in orbit, while one 4000 miles above the surface orbits at only 11,500 miles per hour.

In the preceding discussion we have, for the sake of simplicity, discussed orbits in terms of circular orbits. In general, any satellite orbit is *elliptical,* though a circle is simply one special type of ellipse. An ellipse is another of our conic sections, this one formed by slicing through a cone at some angle with respect to its axis. You may easily draw an ellipse by sticking two pins into your paper, placing a loop of string over them and tracing the figure formed by keeping the string taut with the point of a pencil (Fig. 1.12).

The two pins are at the *foci* of the ellipse you have just drawn. In the case of satellite motion, the orbiting body moves in an ellipse which has the body being orbited at one of its foci. The earth is at one of the foci of the

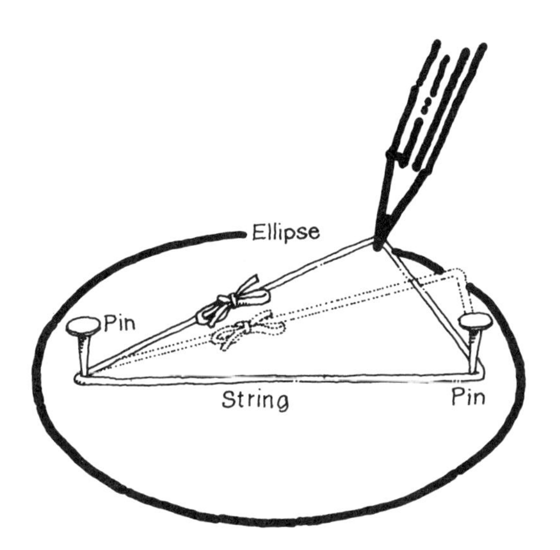

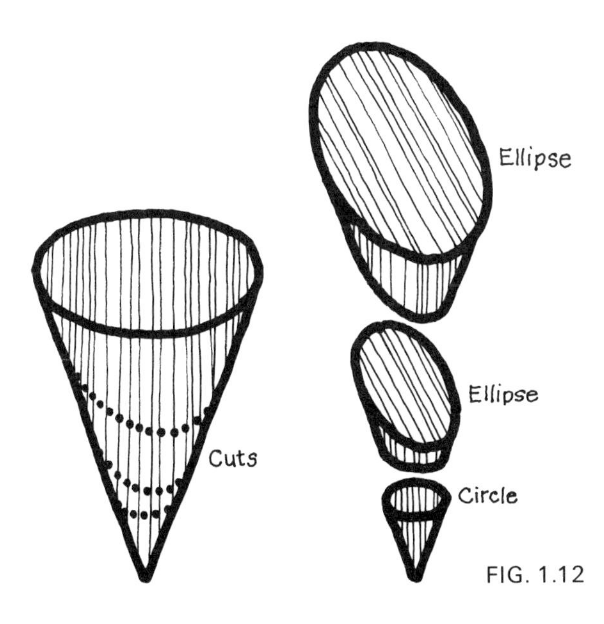

FIG. 1.12

ellipse followed by any earth satellite. In the special case in which the two foci coincide, the ellipse becomes a circle. The positions of a satellite at its greatest and least distances to the center of the earth are called its *apogee* and *perigee*, respectively.

The *eccentricity* of an orbit is a measure of just how elliptical it is. In the case of a circle, for which the apogee and perigee are equal, the eccentricity is zero. An eccentricity equal to a fraction between zero and one denotes an ellipse. The higher the value of the eccentricity, the more elliptical the orbit. When the eccentricity is equal to one, the orbit is no longer closed, but is *parabolic*.

1.6 ESCAPE VELOCITY

We know that any unsupported object near the earth's surface is accelerated downward at 32 feet per second per second. When a ball is tossed straight up into the air, the speed with which it is thrown upward determines how high it will ultimately go. The ball will begin to lose speed as soon as it is thrown and it will slow by 32 feet per second each second until its speed has been reduced to zero. It will then be accelerated back toward the earth at the same rate, so that the upward and downward parts of the motion will be symmetrical: the ball will fall back to the earth with the same speed with which it was thrown upward.

The faster an object is thrown upward, the higher it will ultimately rise. There is a speed great enough that an object thrown upward at that speed will not return to the earth. An object propelled from the earth fast enough not to return is said to have achieved *escape*

FIG. 1.13

velocity from the surface of the earth. The object does not, of course, escape from any influence of the earth's gravitational field; it is simply traveling away from the earth at so great a speed that it will not be slowed to a stop and attracted back to the earth. Just as the motion of a ball tossed into the air is symmetric, an object that is attracted to the earth from a very great distance—such as a meteorite—arrives traveling at the escape velocity.

At the earth's surface, the escape velocity is approximately 7 miles per second, or about 24,100 miles per hour. An object given this speed in an upward direction at the earth's surface would require no further propulsion in order to escape from earth (Fig. 1.13).

As one goes up above the earth's surface, the gravitational interaction is weaker, and hence the required escape velocity is smaller. If an object already in earth orbit is given a speed slightly greater than that which is needed to keep it in orbit at that altitude, it will move into a larger, elliptical orbit. The greater its excess velocity (over that required for a circular orbit at that height) the larger—and more eccentric—the elliptical orbit it will adopt. Finally, if the orbital speed of any satellite is multiplied by the square root of 2 ($\sqrt{2} = 1.414$), its orbit will become parabolic, rather than elliptical, and it will escape. This procedure actually amounts to doubling the energy of the satellite, for any object already in orbit has exactly half enough energy to escape from that orbit. Even though a satellite may escape from its earth orbit, it must still reckon with the gravitational attraction of the sun.

The value of the escape velocity from the surface of any body is a very important factor

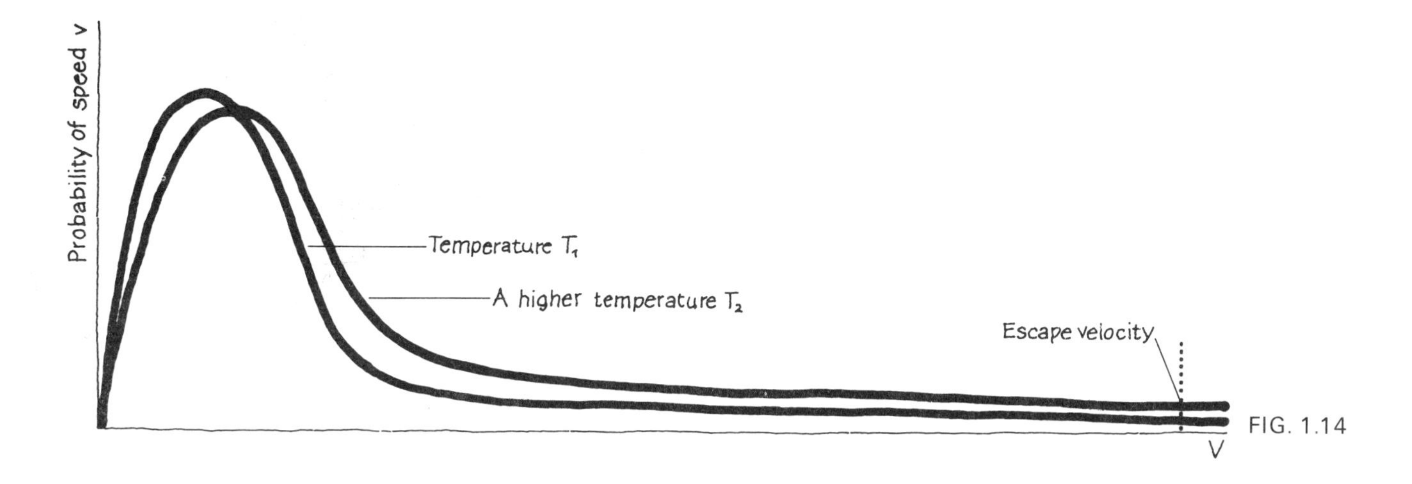

FIG. 1.14

in determining the makeup of the atmosphere that a body can retain, for some of the molecules of its atmosphere may reach escape velocity and be lost to it. The molecules of the gases that make up any atmosphere are in constant motion at speeds dependent on the temperature of the gases. The temperature of any gas, in turn, is simply a measure of the mean *kinetic energy* of the molecules of that gas.

The kinetic energy of an object depends on two things: its mass and its speed. At any given temperature there is a range of speeds which any molecule—nitrogen, for example—may have. Figure 1.14 presents a graph that shows the probability that a nitrogen molecule at a selected temperature will have a given speed. The height of this curve above any given speed (on the horizontal axis) is a measure of the *probability* that any single molecule will be moving at that speed. Expressed in another way, if you had a great many nitrogen molecules at this temperature, the number having any given speed would be proportional to the height of the curve above that speed.

Note that a few of these molecules would be traveling at speeds greater than the escape velocity. This is true of any gas at almost any temperature. Not all those molecules having speeds greater than escape velocity actually escape, for some of them may not be headed in the right direction. But those headed upward at a speed equal to or greater than escape velocity can escape from the earth, provided that they do not collide with another molecule on the way up and become diverted. The dashed curve shows the distribution of molecular speeds for the same gas at a higher temperature. Note that the higher the temperature, the greater the number of molecules that may have speeds greater than escape velocity. This is why a planet like Mercury, with its high surface temperatures, would be expected to have virtually no atmosphere of any sort.

Kinetic energy also depends on the mass of the molecule. At any given temperature, the lighter the molecule the faster it moves. This is true because the molecules of any gas mixture, light and heavy, have the same mean kinetic energies at a given temperature. Since the lighter molecules move faster than the heavy ones at any temperature, more of them have speeds greater than escape velocity. Thus it is easier for light gases to escape than for heavy ones. The lightest gas, hydrogen, is the one that is most easily lost, while carbon dioxide requires much higher planetary temperatures to escape in significant amounts.

The escape velocity from the surface of our moon is only about 1.475 miles per second, about one-sixth that from the surface of the earth. In addition, the moon's surface that faces the sun reaches temperatures as high as 250°F. Whatever atmosphere the moon once had has long since escaped, since its gravitational pull was not sufficient to retain this atmosphere. Escape velocity from the surface of the planet Mars is less than half that from earth, and while Mars has an atmosphere, it is much less dense than ours. In contrast, Jupiter has a very deep atmosphere, because of its combination of high escape velocity and relatively low temperatures.

Part of the Texas–Louisiana Gulf coast, photographed from Gemini XI at an altitude of about 150 miles. The flow patterns of effluents emptying into the Gulf are easily studied from this height. (Photograph courtesy NASA)

1.7 THE TIDES

The gravitational interaction between the earth and its moon affects not only the motion of these bodies through space, but objects on the surfaces of each of these bodies are influenced by the presence of the other body. The gravitational pull of the moon on the earth is easily observed in our system of tides.

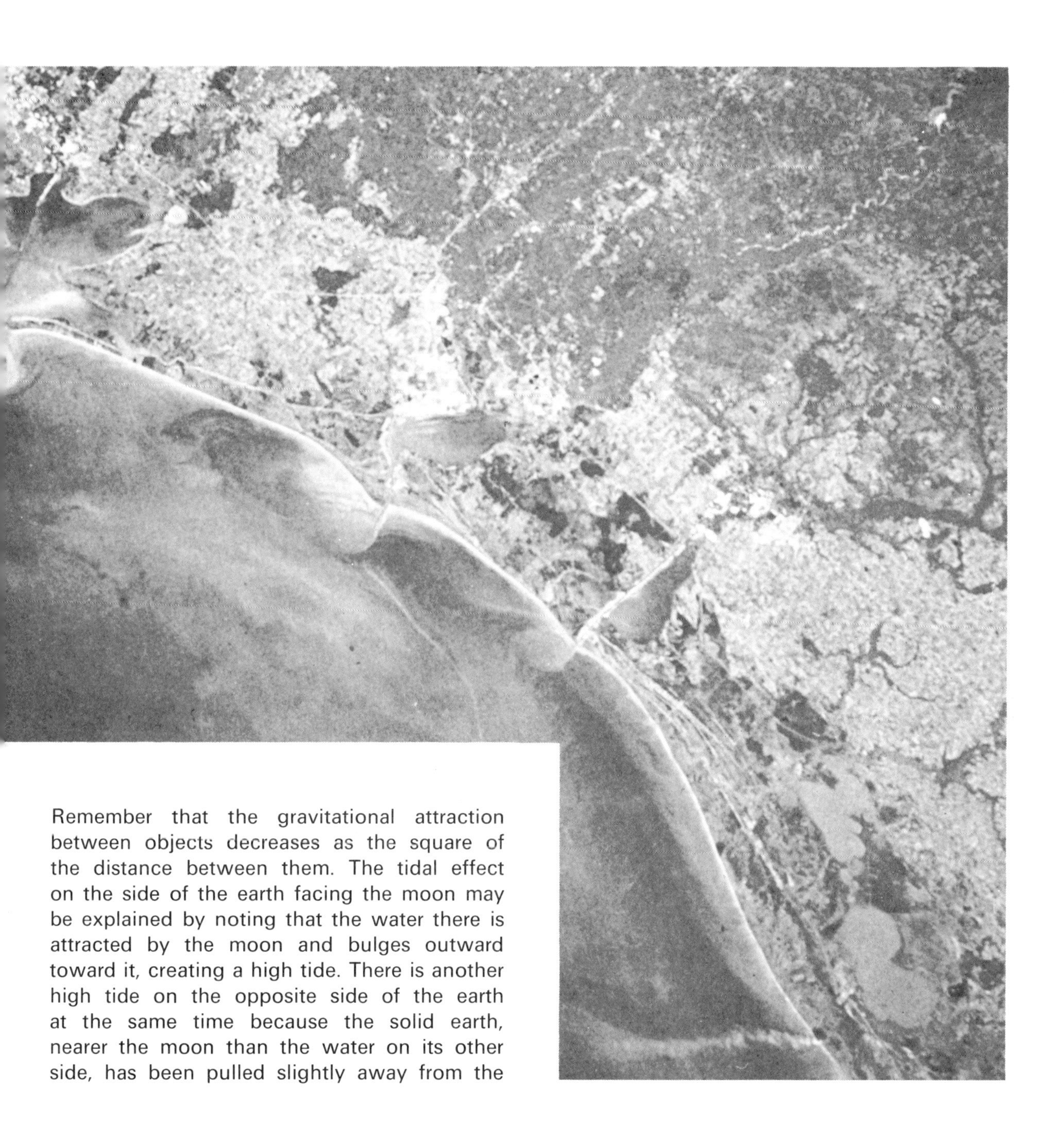

Remember that the gravitational attraction between objects decreases as the square of the distance between them. The tidal effect on the side of the earth facing the moon may be explained by noting that the water there is attracted by the moon and bulges outward toward it, creating a high tide. There is another high tide on the opposite side of the earth at the same time because the solid earth, nearer the moon than the water on its other side, has been pulled slightly away from the

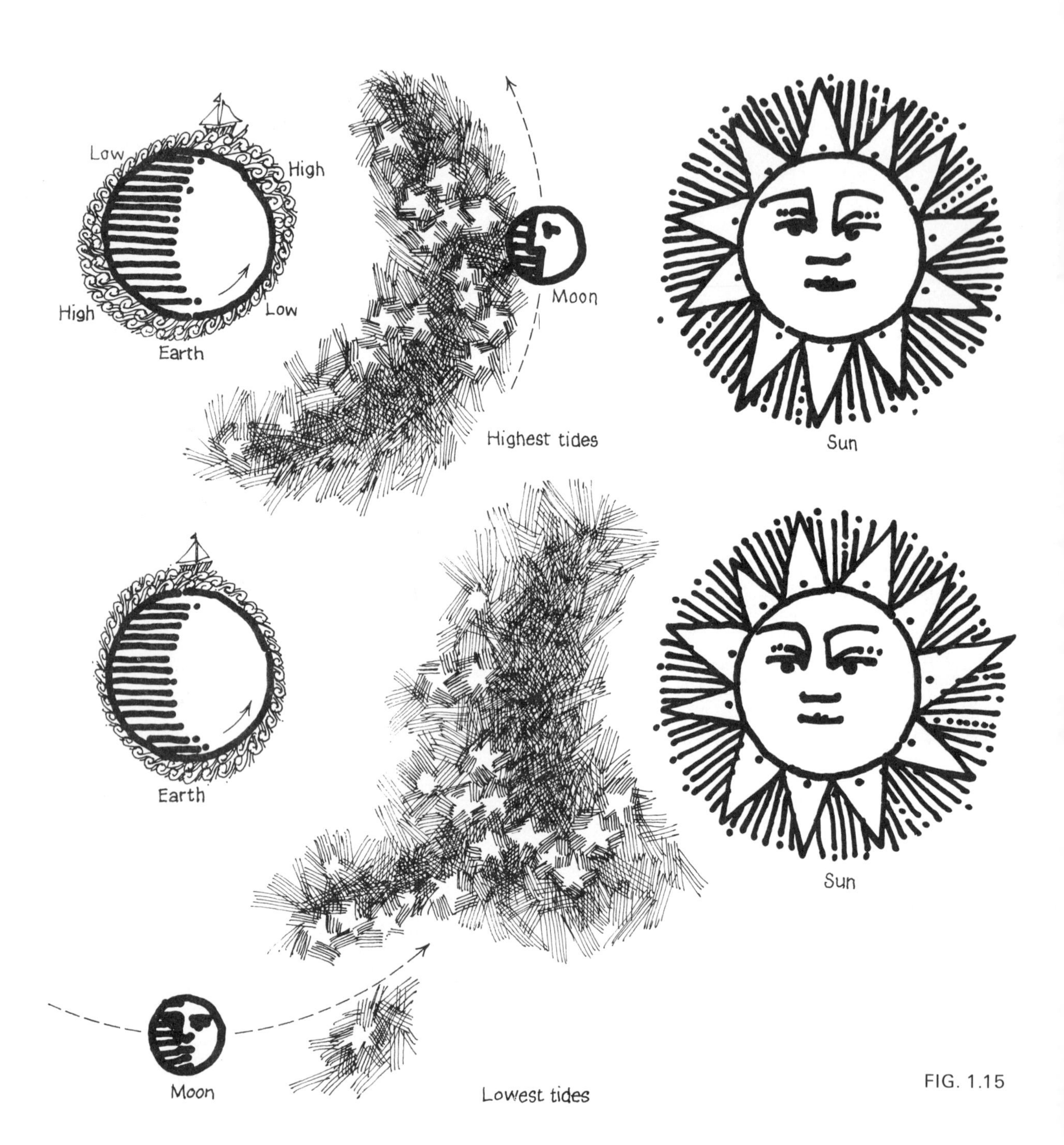
Low
High
High
Low
Earth
Moon
Highest tides
Sun
Earth
Sun
Moon
Lowest tides

FIG. 1.15

water there (Fig. 1.15). The places halfway between these tidal bulges experience, at the same time, a low tide. The position of the high tides is slightly ahead of the moon's position because the earth rotates under these bulges of water.

The extent of the high or low tides depends on the relative positions of the earth, moon, and sun, for the sun's pull upon the earth's oceans causes the same sort of displacement as the pull of the moon. Twice a month the earth, moon and sun are in a line, and then the highest tides are experienced. At other times the sun and the moon partially cancel out each other's effects, producing the lowest, or neap, tides. Of course, tidal peculiarities exist at many coastal points because of the complicating factors of shoreline irregularities, the unevenness of the ocean floor, and other effects. Although not so visible as the ocean tides, movements of the earth's surface due to tidal effects are also observed. When there is a 10-foot tide in the water, the continents rise about six inches.

QUESTIONS

1. What is the weakest force in nature?
2. Is the gravitational force ever observed to be repulsive?
3. If the distance between two objects is doubled, how does the gravitational interaction between them change? What happens if the distance is halved?
4. Discuss the concept of "escape from the earth's gravity."
5. How much would you weigh on the moon?
6. By how much is the speed of an object in free fall increased during each second it falls?
7. Approximately how long does an object have to fall freely to reach a speed of 60 miles per hour?
8. Compare the acceleration due to gravity with that of a car with which you are familiar.
9. Will an object falling from a great height continue to accelerate as it nears the earth?
10. If you fire a gun straight up, will the bullet return to earth with the same speed with which it left the gun barrel?
11. At what angle should one throw a projectile to achieve maximum range?
12. Describe a parabola. How is one constructed?
13. Identify: ICBM, ABM, MIRV, FOBS.
14. Contrast the vertical and horizontal motions of a projectile.
15. Describe the proposed operation of an ABM system.
16. Why can't a satellite remain in earth orbit at an altitude of 10,000 feet?
17. Explain the statement, "The moon is falling."
18. What change in speed is involved if a satellite is moved from a lower into a higher circular earth orbit?
19. Describe the construction of an ellipse.
20. Identify: perigee, apogee.
21. What is escape velocity at the earth's surface?
22. From what has one actually "escaped" when one reaches earth escape velocity?
23. Describe the continual escape of gas molecules from the earth's atmosphere.
24. What planetary conditions are necessary for maximum retention of atmosphere? What conditions will cause a minimum atmosphere to be left?
25. Why does a high tide not coincide with the passage of the moon overhead?

PROBLEMS

1. During the early days of the space programs, the U.S.S.R. was able to put much greater payloads into orbit than the U.S. It was suggested in some quarters that the Russians' secret was a huge, bowl-shaped launch area in which rockets could be carried on rails downhill, accelerated across a relatively level area, then pointed upward at a second hill and launched. Would this scheme have aided the launchings?

2. In Jules Verne's *From the Earth to the Moon,* a space vehicle was fired from earth by a giant cannon. The passengers experienced weightlessness only when passing the point at which the gravitational attractions of the earth and moon cancel each other out. Was Verne's description of this effect accurate? Explain.

3. If you are riding in an automobile that is accelerating very rapidly, you sense this acceleration in a very pronounced way. Yet only a few specially constructed vehicles can sustain an acceleration as great as that of gravity. Do you experience a similar sensation when falling freely?

4. At track and field meets you usually see the shot put or the hammer thrown at about 45-degree angles, but competitors in the long jump take off at about half this angle. Why?

5. Henry was standing on the scales at the rear of an elevator car. As the elevator started upward, he noticed that he had apparently gained 25 pounds, but the scale reading returned to normal when the car stopped. After the elevator had started to move again, Henry noticed that the reading on the scales had dropped to zero. Explain how these things could happen.

6. In one of H. G. Wells' stories, an inventor named Cavor discovered a substance—cavorite—which was repelled by the earth. Cavor proceeded to build a spaceship using cavorite, and was repelled from the earth to the moon. It has long been suggested that if something like cavorite could not be found, at least a "gravity insulator" might be developed. Why would these materials have potentially great usefulness? In what ways might the discovery of one of these materials revolutionize the transportation industry?

7. A synchronous satellite is one which orbits the earth at an altitude such that its orbital period is 24 hours long; thus it remains over the same point on the earth's surface. A suggestion has been made that a long cable be dropped from a synchronous satellite to the earth's surface and this system be used as a "sky hook" to hoist heavy objects into orbit. Comment on this scheme.

8. It may be shown that if a tunnel is cut through the earth between any two surface points, an object will "fall" between the points in about 42 minutes. This would mean that a high-speed transit system might be built between any two cities, with the travel time remaining only 42 minutes, no matter how far apart the cities might be. Discuss the pros and cons of such a transit system between places like Boston, New York, and Washington, or Los Angeles and San Francisco.

9. The United States deployed an ABM system after much debate. Give the arguments that were made on both sides of this issue.

10. If the proverbial "hole to China" were cut through the earth, could you drop an object all the way through?

CHAPTER 2 TRAVEL INTO SPACE

2.1 OUR NEIGHBORHOOD IN THE UNIVERSE

Travel into space has now been removed from the realm of science fiction, and manned missions to the moon are an accomplished fact. Since the problems of manned space flight have been solved, at least in principle, it is possible to theorize about voyages indefinitely far into space. In order to consider where men might go, let's take a look around at our neighborhood in the universe.

Our earth is one of nine objects, known as *planets*, orbiting the star we call our sun, a star that is distinguished from millions of similar stars in the universe only by the fact that it is the one about which our earth revolves. The name "planet" is derived from the Greek word for "wanderer," since it was observed that the planets change their positions in the sky relative to the stars, while the only apparent motion seen in the stars is due to the earth rotating on its axis and orbiting the sun. [The fact that the positions of the stars relative to each other remain fixed over short periods of time (by astronomical standards) makes them extremely useful objects for navigational purposes.]

The planets Mercury, Venus, Mars, Jupiter, and Saturn have been known since ancient times. To early observers, who considered

FIG. 2.1 Ancient idea of the solar system.

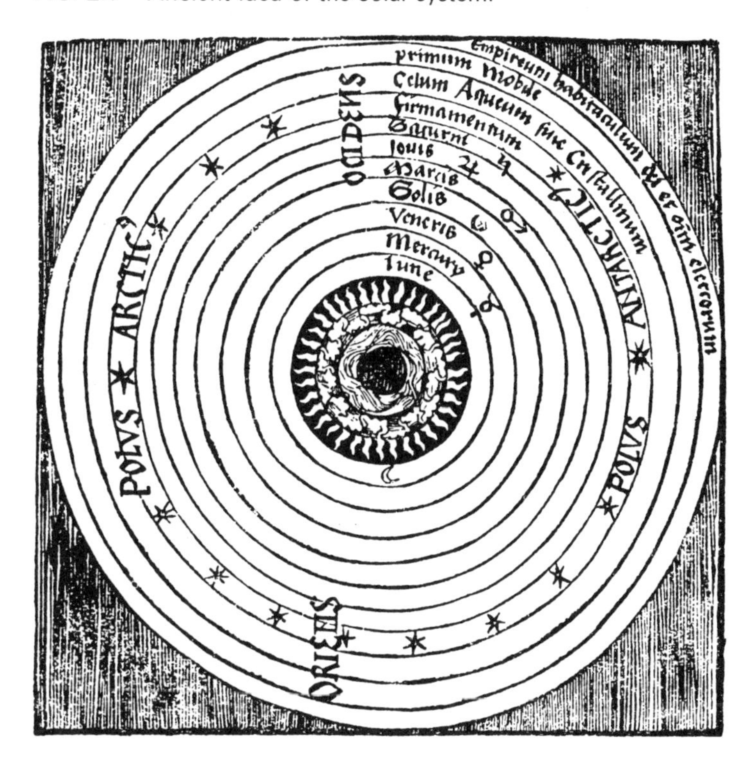

the earth the unmoving center of the universe, the sun and moon were also "wanderers," and a system of seven "planets" is still indicated by the names given to the days of the week (Fig. 2.1). The origins of SUNday,

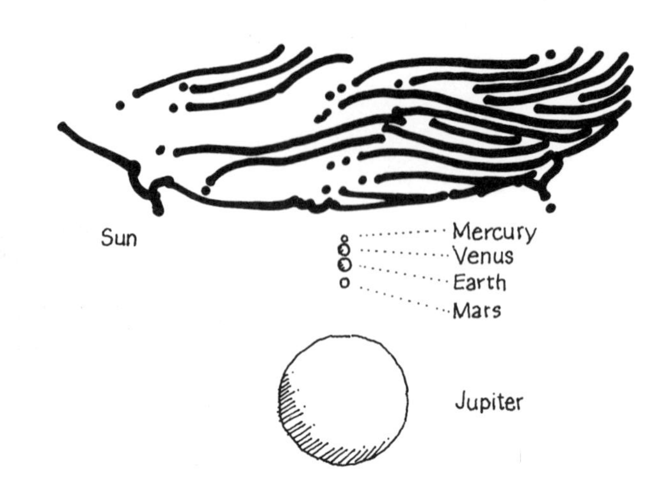

FIG. 2.2 Planets of the solar system.

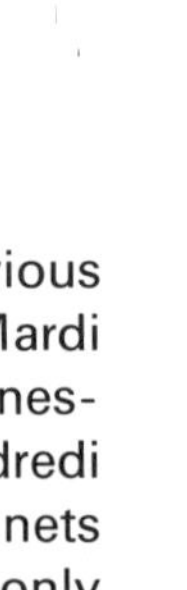

MO(O)Nday, and SATUR(N)day are obvious in English, while in French one finds Mardi (Mars, Tuesday), Mercredi (Mercury, Wednesday), Jeudi (Jupiter, Thursday), and Vendredi (Venus, Friday). The three most distant planets of our solar system were discovered only during the last two centuries, with the aid of telescopes. Uranus was first observed in 1781 and Neptune in 1846, while Pluto was not found until 1930.

All these bodies orbit the sun in the same direction and the orbital planes of all but the innermost and outermost planets lie in virtually the same plane as that of the earth's orbit. The plane in which the earth orbits the sun is known as the *ecliptic*. The orbit of Mercury is inclined at 7 degrees to the ecliptic and that of Pluto is at 17 degrees, while all the other planetary orbits are within three degrees of the ecliptic.

Because of the distances involved in the solar system, it is convenient to define the mean distance from earth to the sun, 92,956,000 miles, as one *Astronomical Unit,* abbreviated

TABLE 2.1 Bode's law for finding distances of planets from sun

Planet	Mercury	Venus	Earth	Mars	|	Jupiter	Saturn	Uranus	Neptune	Pluto
Write 4	4	4	4	4	4	4	4	4	4	
Write 0, 3, 6, ...	0	3	6	12	24	48	96	192	384	
Add	4	7	10	16	28	52	100	196	388	
Divide by 10	0.4	0.7	1.0	1.6	2.8	5.2	10.0	19.6	38.8	
Actual distance	0.39	0.72	1.0	1.52	—	5.2	9.5	19.18	30.07	39.67

AU, and express the distances of the planets from the sun as multiples of this unit (Fig. 2.2). There is a simple rule known as *Bode's law* that will help you remember these distances. It is really incorrect to call this rule a law, for no one knows just why it should give answers that are so nearly correct, and it is not derived from any known physical law.

To use Bode's rule, first write the number 4 in nine columns, one for each of the planets (Table 2.1). Under these numbers write first 0, then 3, 6, 12, ..., doubling each number after the first. Add the numbers in each column, divide by 10, and the results are the approximate distances of the planets from the sun, expressed in astronomical units. The actual values of the planetary distances are recorded, for comparison, below the Bode's-rule distances.

You will notice that there are two apparent exceptions to the rule: there is no known planet at 2.8 AU, and the distance given for Neptune fits Pluto better. However, at about 2.8 AU there *are* located a great number of small bodies known as *asteroids*. The largest asteroid, Ceres, has a diameter of only 470 miles, while others have been observed that are less than a mile in diameter. It has been speculated that a planet once orbited the sun in the space between Mars and Jupiter, but has since exploded, and the asteroids are part of its remains. Alternatively, it has also been suggested that there might have been two planet-sized objects that collided while attempting to establish orbits in this region. The orbit of Pluto actually extends inside that of Neptune, and at times Pluto is nearer the sun than Neptune. This fact, plus the large angle of inclination of Pluto's orbit to the ecliptic, has led to the theory that Pluto is not really a planet but a runaway satellite of Neptune. Apparently there is no imminent danger of a collision between Pluto and Neptune, for a recent calculation has shown that they are locked into a nonintersecting orbital pattern that repeats every 20,000 years.

After one leaves our solar system, the next large object to be found is the nearest star,

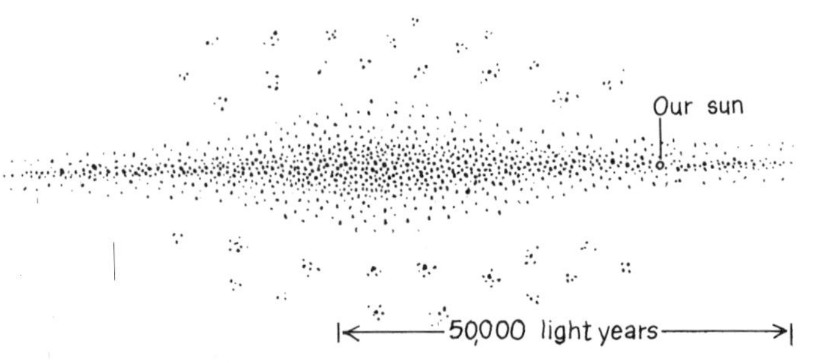

FIG. 2.3(a) Our galaxy

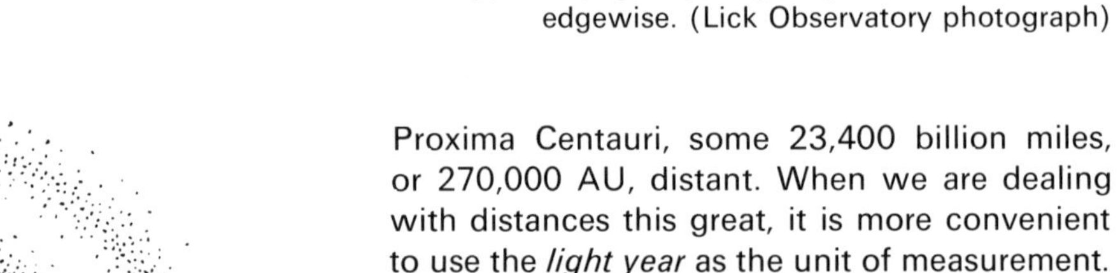

Opposite page: Spiral galaxy NGC 4565, viewed edgewise. (Lick Observatory photograph)

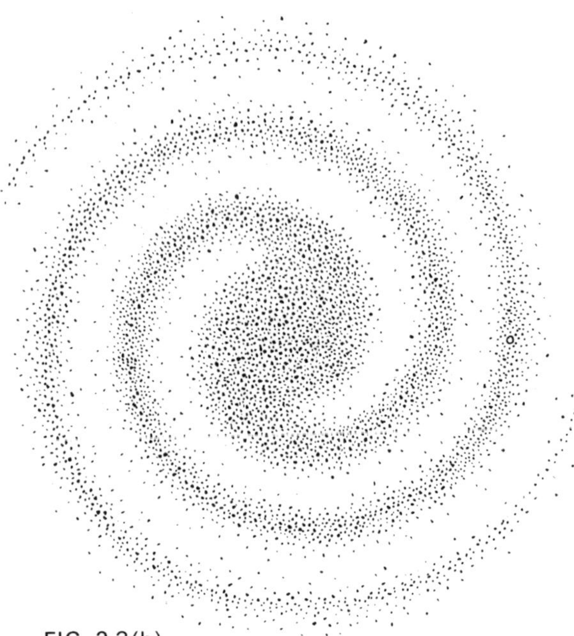

FIG. 2.3(b)

Proxima Centauri, some 23,400 billion miles, or 270,000 AU, distant. When we are dealing with distances this great, it is more convenient to use the *light year* as the unit of measurement. A light year is simply defined as the distance that light will travel in space in one year—about 5880 billion miles. Proxima Centauri is approximately 4.3 light years from earth.

The group of stars among which we live forms our *galaxy*, a collection of stars, interstellar particles, and gases that is shaped roughly like a fried egg (Fig. 2.3a). Our galaxy is known as the Milky Way, and most of the stars visible to the unaided eye belong to it. This entire galaxy is over 100,000 light years in diameter, with a thickness of about 10,000 light years at the center. Our sun is relatively near the edge of the galaxy, about 30,000 light years from its center. Viewed from above, the stars

of the Milky Way form a spiral (Fig. 2.3b). The spiral arms rotate about the center and our sun is moving at about 12 miles per second with respect to the galactic center. Ours is only one of some 100 billion galaxies located within 10 billion light years from earth.

2.2 VEHICLES FOR SPACE TRAVEL

Now that we've had a brief look at the "road map" of the space about us, let us consider the sort of vehicles in which one may travel through space. Since there is a fairly good vacuum there (the atmospheric pressure is less than one-billionth of the atmospheric pressure near the earth's surface), any space vehicle should be sealed, carry its own environmental system with supporting facilities, maintain on-board navigation and communications equipment, and have power sources to operate all its equipment. Nothing yet mentioned presents any problem that has not yet been solved, though on a different scale, for the operation of high-altitude commercial jet aircraft, for the environment at 35,000 feet above the earth's surface is as deadly to a human without life-support systems as that 300 miles above the earth. Of course aircraft use air-breathing engines which cannot be used in space, are aloft for relatively short periods at a time, and travel much more slowly than is desirable for space missions.

Space vehicles must depend on reaction engines for propulsion. A balloon that has been blown up and then released with its mouth open is an example of the simplest sort of reaction propulsion device. The balloon carries its

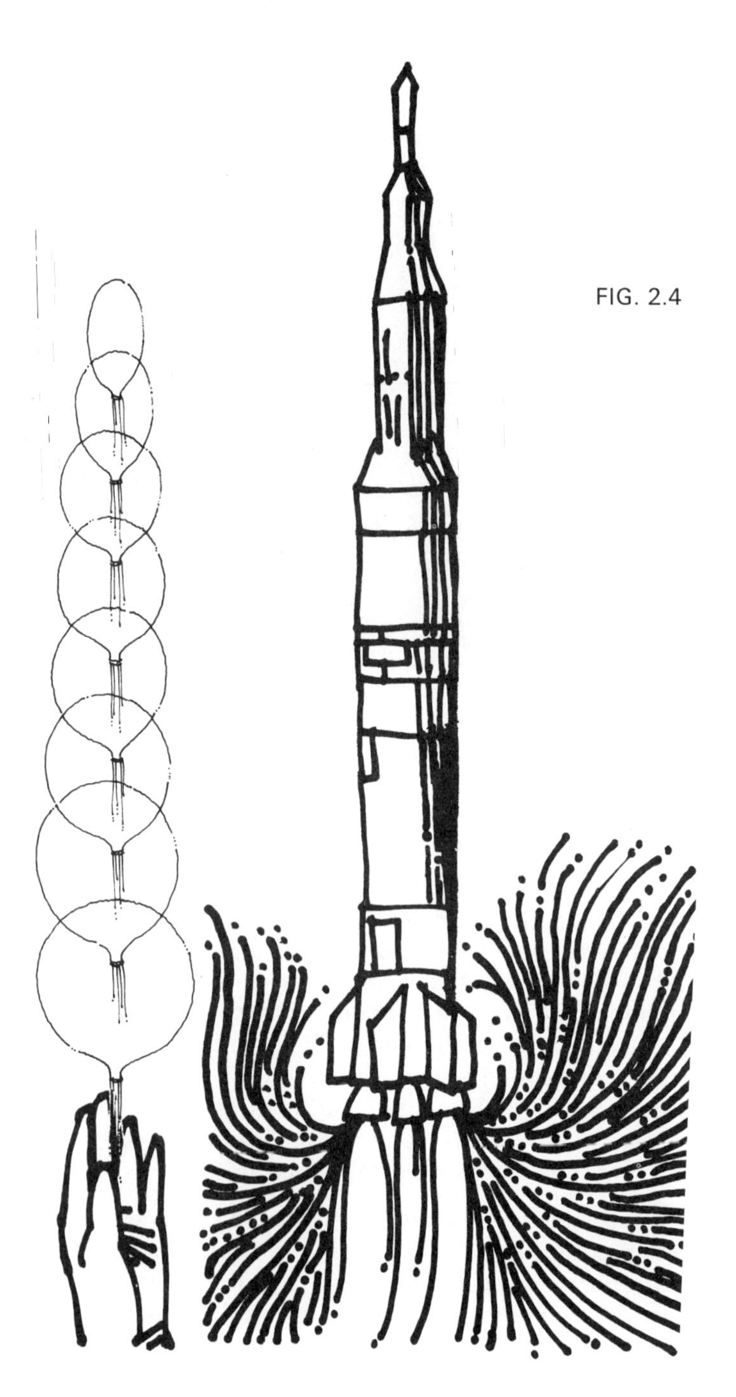

FIG. 2.4

propellant, the air compressed inside, and this air exhausting from behind drives the balloon forward (Fig. 2.4). A rocket engine carries its fuel and oxidizer, the two are combined and ignited, and the hot exhaust gas rushing from the engine ports causes the rocket to be propelled forward. This exhaust gas, in propelling the rocket, does not push against anything, for the gas moving rapidly in one direction is sufficient to cause the rocket to move in the opposite direction.

A physical way to describe this process is to say that the rocket engine and its fuel make up an isolated system in which the *momentum* must remain constant at all times. The momentum of any object is defined as the *product* of its mass and its velocity. Thus a light but rapidly moving object may have the same

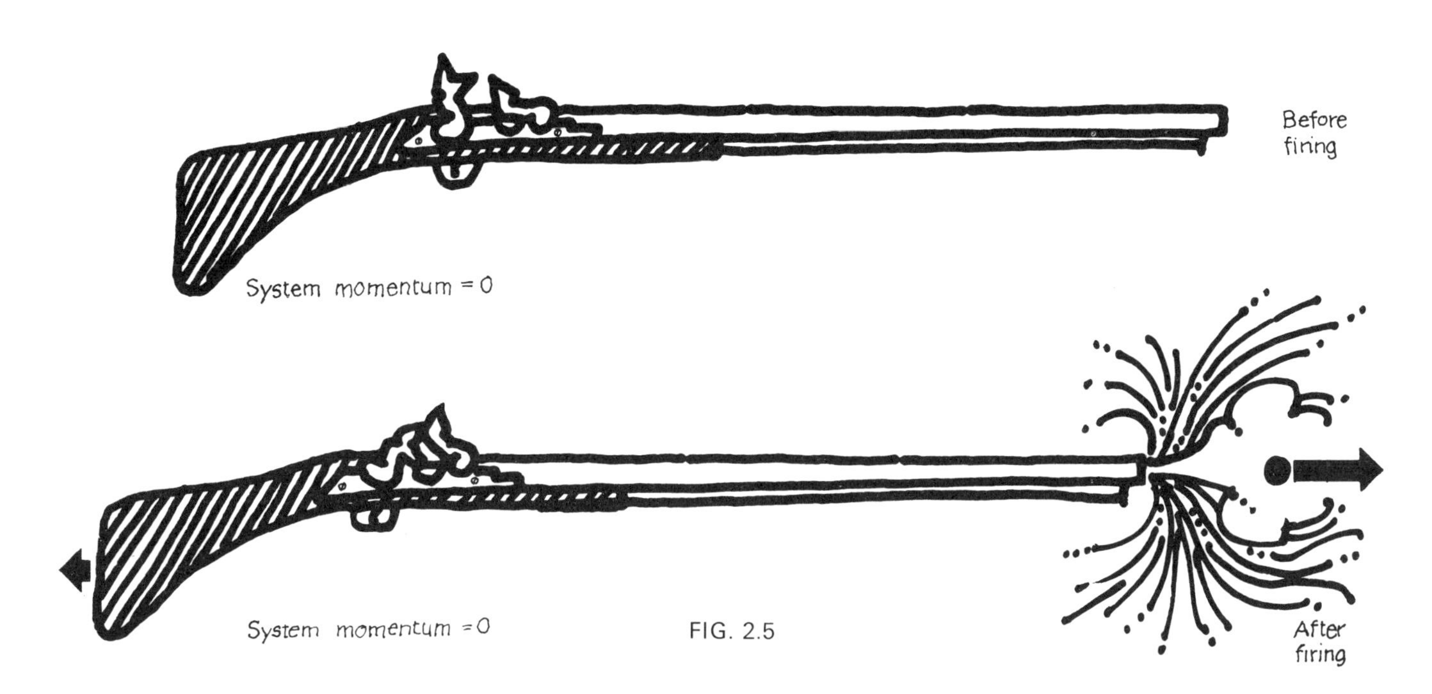

FIG. 2.5

magnitude of momentum as a heavy object moving slowly. In any isolated system (isolated in the sense that the system has no net external force acting on it), the total momentum must remain constant, i.e., it must be *conserved.*

Anyone who has ever fired a gun has observed an example of conservation of momentum. Before the gun is fired, both it and the bullet in the firing chamber are at rest and the total momentum of the system (consisting of the gun plus the bullet) is therefore zero. Immediately after the gun is fired, the bullet travels in one direction at a high speed; thus there is now a momentum in the forward direction due to the motion of the bullet. In order to conserve momentum, the bullet's momentum (bullet mass times bullet velocity) has to be exactly matched by the gun's momentum (gun mass times gun velocity) in the backward direction (Fig. 2.5). Because a gun may be many times as massive as the bullet it fires, its velocity may be only a fraction of that of the bullet, yet no one who has ever fired a rifle without properly supporting it against his shoulder can doubt that the gun moves backward immediately after being fired. [Of course, a gun is normally supported when fired, and any support that is used becomes a part of the system (gun plus bullet plus support). To make the backward velocity of a gun after firing as small as possible, one adds as much mass as possible to the system. If a gun is held firmly against one's shoulder, the mass of the person firing the gun is added to the system. Still more mass may be added to the system if the shooter braces himself against a tree or lies on the ground.]

A rocket engine works because fuel is burned and exhausted at high speed. In order for the momentum of the system (rocket plus fuel) to be conserved, the vehicle has to speed up as the fuel is exhausted, and the speed of the rocket-propelled vehicle will continue to increase so long as fuel continues to be burned and exhausted. The actual burning of the fuel is of only secondary importance in the operation of the rocket, for it would operate just as well if some other means could be devised to cause the fuel to be exhausted from the engine at a very high speed. Reaction propulsion engines that operate by means other than the combustion of chemical fuels are currently the subject of research for deep-space usage.

2.3 SPACE VEHICLE DESIGN

To better understand what goes into any space vehicle, let's consider some things basic to their design. There are three inherent parts of any space vehicle:

1. *The payload* (this is simply what you want to carry and deliver somewhere, plus the support systems required by the payload).
2. *The superstructure* (the shell and framework of the ship itself).
3. *The fuel* (the expendable supply of fuel carried for propulsion purposes).

At the time of its launching, the mass of the space vehicle will simply be the sum of the three components listed above. After the fuel has all been expended, the vehicle's mass will equal only the sum of the first two items. In considering the performance of any reaction propulsion ship, it is convenient to define the *mass ratio* of the ship:

$$\text{Mass ratio} = \frac{\text{mass of vehicle as it sits on launch pad}}{\text{mass of vehicle after all fuel has been burned}}$$

The mass ratio is obviously the sum of all three items in the list above, divided by the sum of items 1 and 2.

We may use the mass ratio to calculate the final speed of the ship after all the fuel has been burned. No matter how fast the ship is moving with respect to the ground, or any other object, the fuel leaves the ship with the same speed *with respect to the ship*. Thus the space vehicle's speed with respect to a fixed observer may exceed the speed with which the fuel is exhausted from the ship. The approximate relationship between mass ratio and vehicle speed is indicated in Table 2.2.

TABLE 2.2

Final vehicle speed / Fuel exhaust speed	Mass ratio of ship
1.0	2.72
2.0	7.40
3.0	20.0
4.0	52.0

We can readily see that the mass ratio must increase enormously if the final speed of the vehicle is to be many times that of the fuel exhaust speed. The limiting factors for the mass ratio are simply the structural strength of the vehicle and the inherent strengths of the building materials. The exhaust velocity of currently available chemical fuels is about 6700 miles per hour. To reach escape velocity from the surface of the earth with a single-stage rocket requires a mass ratio of about 50. This ratio is above the limit we can expect from present building materials and techniques; hence multistage vehicles are required for most space travel.

The Apollo series of spacecraft that carried the first men to the moon are propelled by typical multistage rocket boosters. The first of the three stages lifts the ship to an altitude of about 40 miles and, having burned all its fuel, is jettisoned. At this time the vehicle is traveling at about 5300 miles per hour. The second-stage engines accelerate the ship to about 15,300 miles per hour and carry it to more than 100 miles above the earth's surface, where *it* is released. The third-stage engine powers the ship into an earth orbit, at about 17,500 miles per hour, and is later restarted to drive the craft out of earth orbit and on toward the moon.

2.4 THE LIMIT OF OUR TRAVEL INTO SPACE

Within the limitations of present technology, a final vehicle speed of about 25,000 miles per hour is attainable. Since we always expect technological advances for the purposes of our considerations, let us suppose that speeds of 100,000 miles per hour are attainable in the near future. At such speeds, how far can we expect to go into space within a reasonable period of time?

The nearest star, Proxima Centauri, is about 4 light years away. A trip to this star at a speed of 100,000 miles per hour would require 25,000 years! (One way, that is.) Obviously, in terms of our present propulsion capabilities, travel even to this nearest star is out of the question.

However, alternatives to chemically fueled rockets are now under active consideration. *Nuclear propulsion systems,* currently being developed, use the heat from a nuclear reactor to raise liquid hydrogen to very high temperatures. The molecules of the hydrogen gas, moving at great speeds, exhaust from the engine, driving the ship forward. One great advantage of a nuclear propulsion engine is that it may have very high exhaust velocities. Disadvantages include radiation hazards, the weight of the reactor, and its necessary shielding. Nuclear propulsion engines, which are never intended to be used in the earth's atmosphere, will probably power the first manned mission to Mars.

Still greater exhaust velocities may be obtained from propulsion devices that accelerate charged particles. These *ion propulsion engines* would also be useful only in deep space. The limiting speed for exhaust particles, whatever the type of engine developed, is the speed of light. Therefore the use of light beams—streams of photons—has been suggested as a power system. A *photon propulsion system* would accelerate a vehicle very slowly, but might be operated for long periods of time during long space voyages.

Perhaps the ultimate propulsion system that has been suggested would use exploding thermonuclear bombs to drive a space vehicle. A possible H-bomb-powered ship is illustrated

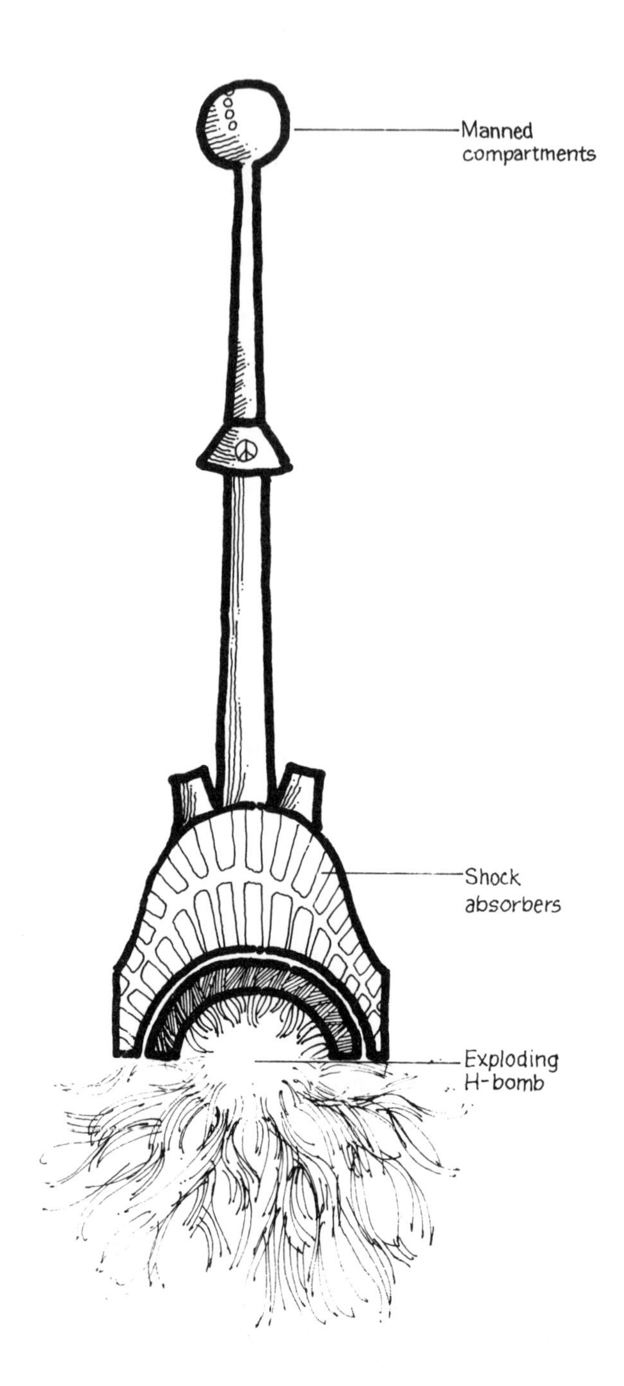

in Fig. 2.6. One study of such a system suggests that an ultimate speed of 2,000,000 miles per hour would be attainable. But even at this unthinkable speed, the time it would take to travel to the nearest star would still be almost 1000 years!

It is, of course, possible that small, self-contained, self-sustaining colonies might start on deep-space journeys fully aware that those persons who set out from earth would not live to see even their destination in space and only their distant progeny would be able to eventually return to earth. But the many human problems involved probably preclude this sort of mission. An alternative means of extending human capability for deep-space travel involves lengthening the amount of time the traveler has available for the trip. Phenomena involving either *relativity* or *cryogenics* (see below) have been viewed as offering possible means of increasing the effective human life span.

In Chapter 3 we shall discuss the relativistic *time-dilation* effect, that is, the fact that time

FIG. 2.6 Proposal for an H-bomb-powered space vehicle.

appears to pass more slowly for persons in motion at very great speeds than it does for stationary observers. At a sufficiently high speed, a space traveler might experience the passage of only a year, as he measured the time, during the same period in which persons on earth experience the passing of ten or more years. Unfortunately, this effect becomes significant only when the traveler is moving very near the speed of light, and no known or proposed propulsion system is capable of delivering the speeds required.

The possibility of cooling the bodies of space travelers to cryogenic (very low) temperatures, thus slowing down or even temporarily stopping their life processes, has also been suggested. In the discussion in Chapter 10, we shall see that no satisfactory means of cooling and subsequently reviving even small mammals exists at present, though research in this area continues.

Travel within our own solar system, however, is a bit more realistic. Even the most distant planet, Pluto, *could* be reached by a direct flight in less than ten years, using only conventional rockets. It must be pointed out, however, that space travel is not always a simple straight-line-path affair. Unlike television's "Star-Trek" type of voyage, in which the engines of the space vehicle could apparently operate continually, chemically fueled rockets cannot carry enough fuel to operate except in bursts. Thus a present-day space vehicle is accelerated during a very short time interval to a speed sufficient to reach its objective, and thereafter, except for possible midcourse corrections, is left to "fall" in the gravitational fields of the sun and any nearby massive objects. In Section 1.6 we noted that even an object that has reached earth escape velocity is still under the gravitational influence of the sun. For this reason, travel within the solar system primarily involves transferring from one solar orbit (the earth's) to the orbit of the body one wishes to reach, and return.

Figure 2.7 shows typical trips which have been made by unmanned probes to the two nearest

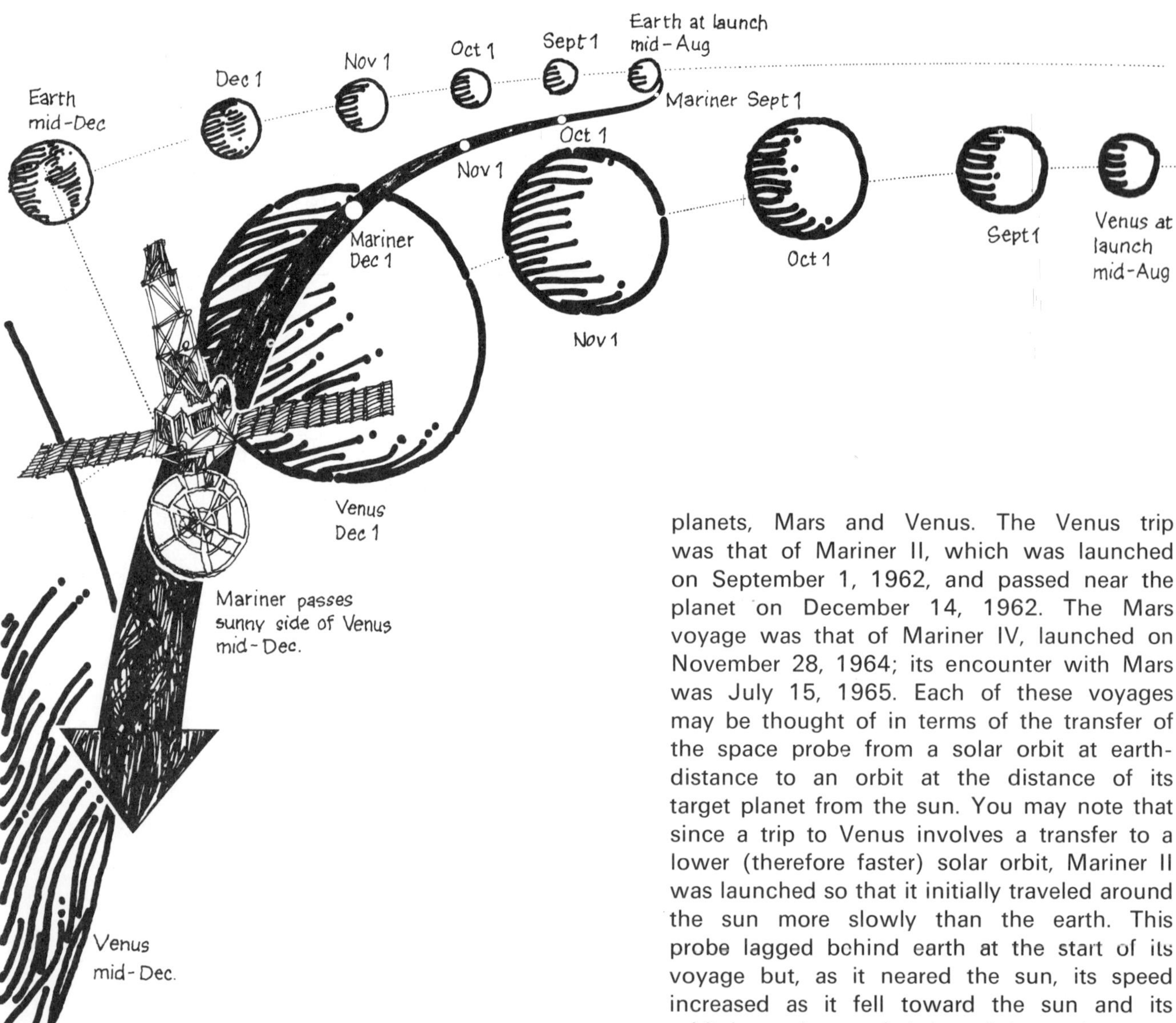

FIG. 2.7(a) Mariner II trajectory.

planets, Mars and Venus. The Venus trip was that of Mariner II, which was launched on September 1, 1962, and passed near the planet on December 14, 1962. The Mars voyage was that of Mariner IV, launched on November 28, 1964; its encounter with Mars was July 15, 1965. Each of these voyages may be thought of in terms of the transfer of the space probe from a solar orbit at earth-distance to an orbit at the distance of its target planet from the sun. You may note that since a trip to Venus involves a transfer to a lower (therefore faster) solar orbit, Mariner II was launched so that it initially traveled around the sun more slowly than the earth. This probe lagged behind earth at the start of its voyage but, as it neared the sun, its speed increased as it fell toward the sun and its orbital speed exceeded that of the earth.

Conversely, the Mars probe was launched at a greater speed than that of the earth with respect to the sun. Thus as the probe flew

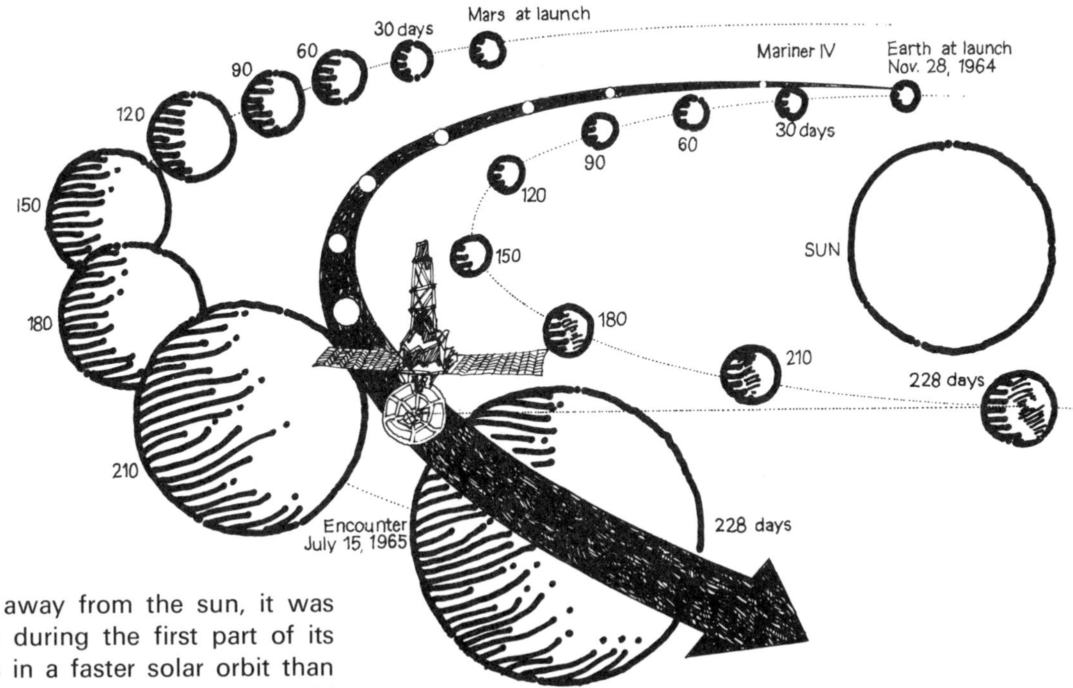

FIG. 2.7(b) Mariner IV trajectory.

out of earth orbit, away from the sun, it was ahead of the earth during the first part of its trip. The earth was in a faster solar orbit than the probe, however, and it passed Mariner IV before the probe reached Mars.

Travel time from earth to the distant planets may be significantly reduced by using the gravitational pull of the other planets, particularly Jupiter, to speed up the space vehicle as it flies by. The operation of this swing-by technique is illustrated in Fig. 2.8. *As seen from the planet,* the spacecraft approaches and leaves on a hyperbolic orbit, with its incoming and outgoing speeds equal at equal distances from the planet. But the speed of the spacecraft *relative to the sun* is changed (the planet is moving with respect to the sun, of course) and the spacecraft can leave Jupiter flying faster than it was when it approached. The use of one or more of the planets in this manner can greatly reduce the time needed to travel in the solar system.

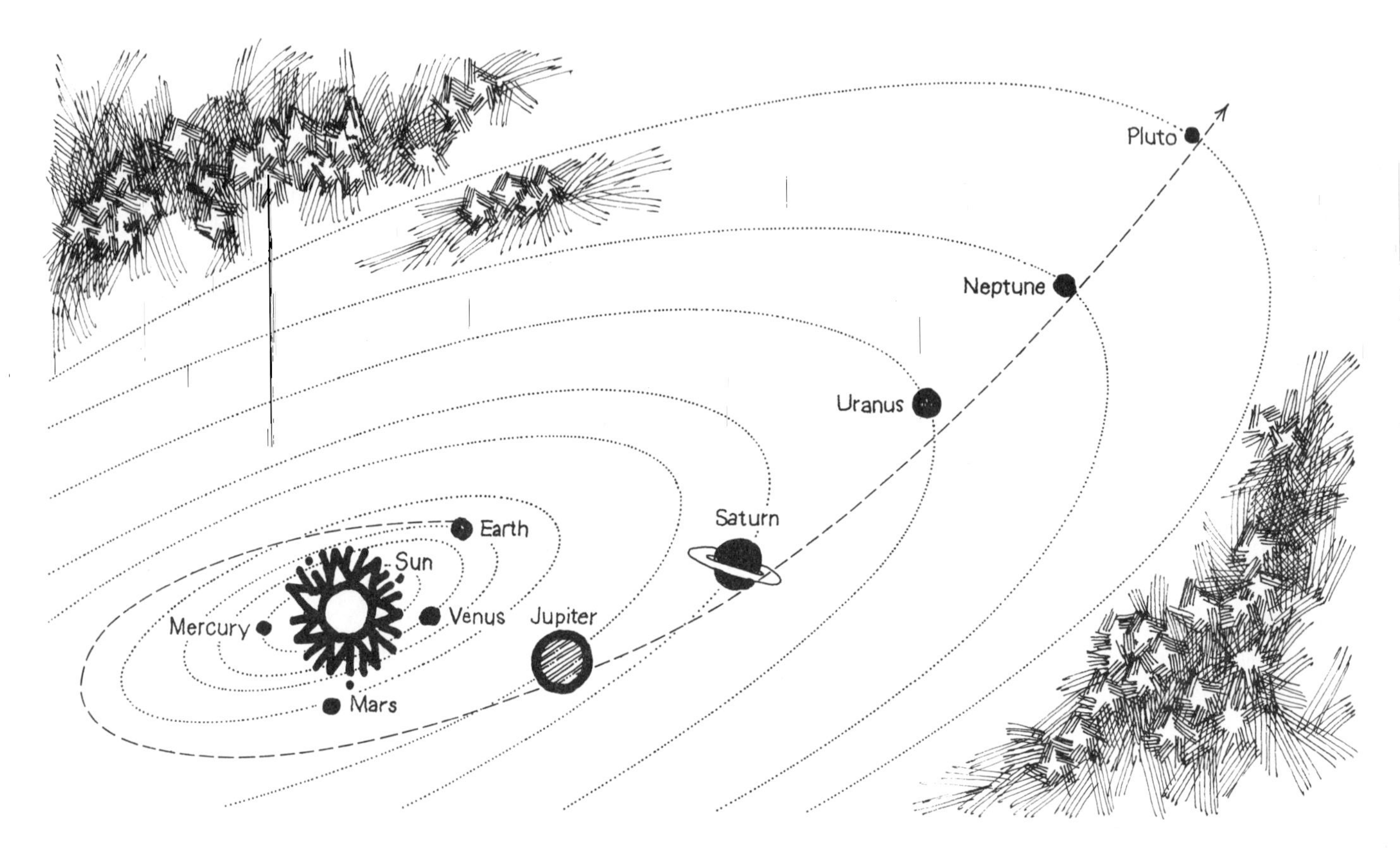

FIG. 2.8 "Grand tour" of Jupiter, Saturn, Uranus, and Neptune.

Between 1976 and 1980 the outer planets will be in an unusually favorable alignment for swing-by missions. For example, a vehicle launched during September, 1977, could use the help of Jupiter and Saturn to reach Pluto in only $8\frac{1}{2}$ years, while a direct flight to Pluto would require more than 40 years using the same vehicle propulsion system but without use of the swing-by technique. During this same period, a planetary alignment that occurs only every 175 years will permit a Grand Tour of the outer planets, with fly-bys of Jupiter, Saturn, Uranus, and Neptune all being possible in only eight years.

2.5 OUR SOLAR SYSTEM: WHAT CAN WE EXPECT TO FIND IN IT?

Since our solar system appears to be the extent of the part of space accessible to us for exploration within the near future, we shall now consider what we already know about it and ask some of the questions to which we hope

TABLE 2.3

Planet	Symbol	Mean distance from sun, AU	Period of solar orbit	Inclination of orbit, °	Eccentricity of orbit	Mean diameter, miles
Mercury	☿	0.39	88 days	7	0.206	3 025
Venus	♀	0.72	225 days	3.4	0.007	7 526
Earth	⊕	1.00	1 year	—	0.017	7 926
Mars	♂	1.52	1.9 years	1.8	0.093	4 180
Jupiter	♃	5.20	11.9 years	1.3	0.048	88 700
Saturn	♄	9.54	29.5 years	2.5	0.056	75 000
Uranus	♅	19.18	84.0 years	0.8	0.047	29 600
Neptune	♆	30.07	164.8 years	1.8	0.009	27 600
Pluto	♇	39.57	247.7 years	17.2	0.249	3 600?

Planet	Period of revolution	Mass (relative to earth)	Acceleration due to gravity at surface (relative to earth)	Escape velocity, miles/second	Number of satellites
Mercury	59 days	0.05	0.39	2.5	0
Venus	243 days	0.81	0.87	6.2	0
Earth	23 hr 56 min .04 sec	1.00	1.00	6.95	1
Mars	24 hr 37 min	0.11	0.38	3.1	2
Jupiter	9 hr 50 min	317.4	2.64	38.	12
Saturn	10 hr 14 min	95.1	1.16	23.	9
Uranus	10 hr 42 min	14.6	1.02	13.7	5
Neptune	15 hr 48 min	17.2	1.34	15.5	2
Pluto	153 hr	?	?	?	0

to find answers. A summary of some important facts about the members of the solar system is contained in Table 2.3.

The Earth's Moon

The extraterrestrial body closest to the earth—its moon—is the natural starting point for our explorations into space. At an average distance of only 238,856 miles from earth, the moon has been the object of numerous earth-based observations, but until we were able to land on it and bring small pieces of it back to our laboratories for analysis, we really knew little about our nearest neighbor. We did know that the moon's diameter is about 2161 miles, its average density is three and one-third times that of water, and the acceleration due to gravity at its surface is only one-sixth that at the earth's surface. The moon has no atmosphere or free water and its surface temperature varies between about 220°F and −240°F during the 27-day-long lunar "day."

A portion of the moon's surface, photographed from Apollo 8 (December 1968). The crater at the bottom is Goclenius, unusual for the prominent rille that crosses the crater rim. (Photograph courtesy of NASA)

Although we consider the moon as earth's satellite, in another sense it may be regarded as a small planet that happens to orbit the sun at about the same distance from the sun as the earth does. To accurately describe the motion of the earth–moon system about the sun, one should think of an asymmetric, spinning dumbbell whose center of mass executes a near-circular orbit about the sun. Properly, the moon does not orbit the earth any more than the earth orbits the moon, but each of them orbits their common center of mass. This center of mass (which would be the balance point of our dumbbell in a uniform gravitational field) happens to be located beneath the earth's surface, 2900 miles from the center of the earth, because the earth is about 81 times as massive as the moon.

The moon always keeps the same face toward the earth, so one of the first things we were interested to learn from space exploration is that the hidden side of the moon looks somewhat different from the side we can see. There are more craters on the far side of the moon than on the near side, and the large, smooth-appearing areas seen on the moon's face are virtually absent from the back. From earth these large areas appear to be relatively smooth, so ancient observers named them *maria*, or seas. But a closer look at the moon's surface has revealed that it is incredibly rough and even the maria are pitted with craters everywhere. The density of the material of which the maria are made is greater than that of surrounding areas, and spacecraft in lunar orbit therefore experience gravitational irregularities while passing over them. The MASs CONcentrations responsible for these orbit perturbations are known as *mascons.*

At 4:05 P.M., EDT, on July 20, 1969, two American spacemen, Neil A. Armstrong and Edwin E. Aldrin, landed on the Sea of Tranquillity and thereby became the first earth men to touch the moon. They returned to earth with samples of lunar rocks and soil and left several pieces of scientific apparatus on the moon's surface. The success of their Apollo 11 mission marked the beginning of a new era in extraterrestrial exploration and investigation. A second mission, Apollo 12, landed on the Ocean of Storms on November 19, 1969, and also returned with rock samples.

The lunar surface at Tranquillity Base was described as "slippery" by the astronauts who walked on it, a fact that is hardly surprising now that it has been found that this soil is a mixture consisting mainly of small glass and crystalline fragments. Many of the rocks gathered at the landing site were found to be *igneous*—formed under intense heat—giving us a clue to part of the moon's past. The ages of the rock samples taken from both the Sea of Tranquillity and Ocean of Storms sites were measured to be between 3.3 and 3.7 million years, except for one highly radioactive rock, from Apollo 12, which has been found to be 4.6 billion years old. This rock apparently dates back to the formation of the solar system!

All the moon rock samples collected during the Apollo 11 and 12 missions have unusually high concentrations of titanium, scandium, zirconium, and some rare-earth elements, though the igneous rocks are surprisingly deficient in europium (also a rare-earth element). The lunar soil was found to be rich in nickel, cadmium, zinc, gold, silver, and copper. Some

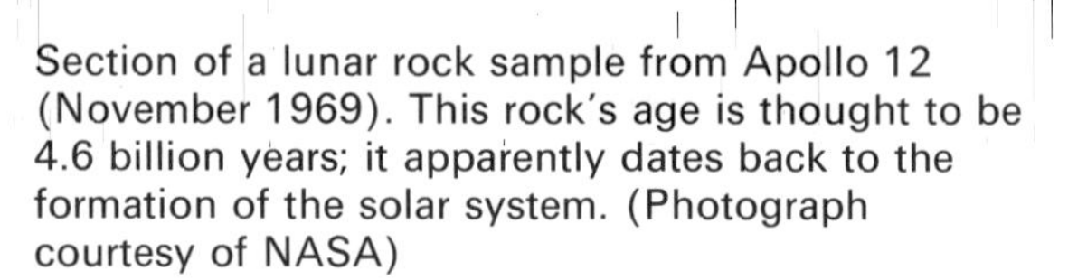

Section of a lunar rock sample from Apollo 12 (November 1969). This rock's age is thought to be 4.6 billion years; it apparently dates back to the formation of the solar system. (Photograph courtesy of NASA)

of the rock samples contained as much as 7% titanium. Studies of the magnetic properties of lunar materials indicate that the moon either had a strong magnetic field at one time or it was once close to an object that had such a field.

Studies of rock and soil samples collected from these two small areas on the moon's surface have already dashed any hopes that the moon will easily reveal its secrets. Despite the fact that the sites thus far visited were both maria areas, the rock samples from the two locations differ dramatically. About half the Apollo 11 samples were *breccias*—rocks fused together by meteor impacts—but breccias were almost totally absent from the Apollo 12 samples. The 4.6-billion-year-old rock was found to contain 20 times as much uranium, thorium, and potassium as the other samples, and is apparently representative of a still different family of lunar materials.

Seismometers—instruments able to detect moonquakes, meteor impacts, or possible volcanic activity—were left on the moon's surface by each Apollo mission. These instruments indicated that the moon is relatively "quiet," seismically. However, when the Apollo 12 lunar ascent module was sent crashing into the surface, the seismometers showed that the moon "rang like a gong" for forty minutes afterward. The somewhat heavier third-stage booster of the aborted Apollo 13 mission produced a similar effect, as the moon shuddered for four hours after its impact. While the structure of the interior of the moon remains unknown, it is certain that it is very different from that of the earth.

A *retrodirective reflector*—a reflector that reflects light incident upon it back in the direction from which it came—that measured about 18 inches square was placed on the moon's surface by the Apollo 11 crew. Since then,

Apollo 11 on the moon, with Edwin Aldrin in foreground, standing next to the seismometer. Beyond the seismometer is the retrodirective reflector. (Photograph courtesy of NASA)

earth-based lasers, aimed through telescopes, have been bounced off this reflector and their reflections detected on earth. With the aid of this reflector, highly accurate measurements of changes in the moon's orbit and the earth–moon distance will be possible.

These first expeditions to an alien body have demonstrated man's ability to survive and function on planets other than earth. The question, "Can men ever live on the moon?" has been answered, for men already *have lived* on the moon. It is true that human life on the moon has thus far been of extremely short duration and has been accomplished only by bringing from earth all supplies needed to sustain life. However, under laboratory conditions, earth plants have been grown successfully in samples of lunar soil. It is not difficult to envision that someday crops growing on the surface of the moon will provide oxygen, food, and water to earth men living there.

Venus

Turning our attention to the rest of the solar system, we are perhaps most interested in the two planets that are nearest to and most like our earth: Venus and Mars. Venus has been called the earth's twin because of the many similarities between the two. They are nearly the same size: the diameter of Venus is about 7526 miles, while that of the earth is 7926 miles, and the acceleration due to gravity at the surface of Venus is about 87% of that on earth. Venus' orbit of the sun requires 224.7 earth days as it moves in a nearly circular path at a distance of about 67,000,000 miles from the sun.

Since Venus has been known from ancient times, many observations of its surface have been attempted from earth-based stations. Making such observations is by no means easy, since it is impossible for an observer on earth to look at Venus with the sun behind him. Venus appears as the legendary Morning and Evening Stars, and observations made during darkness find the planet low in the sky, enabling one to view it only through a thick blanket of the earth's atmosphere.

The surface of Venus is obscured by a thick cloud cover that is highly reflective. Because so much of the incident solar radiation is reflected, it was once thought that the surface temperature of Venus is near that of the earth. One of the first clues to the makeup of the Venusian atmosphere came when it was discovered that photographs made with film sensitive to ultraviolet light revealed a great deal of detail below the clouds. This fact ruled out the possibility of there being significant amounts of water in the dense cloud cover, since water vapor strongly absorbs ultraviolet radiation.

Another long-standing mystery was the period of rotation of the planet on its axis. It was once thought that the Venusian day should be nearly the same length as an earth day, since a day on Mars lasts only about half an hour longer than on earth. Although the surface of Venus still cannot be visually observed from earth, it is possible to bounce radar signals off the planet and the echoes of these signals indicate that a single revolution takes 243 days, which is *longer* than it takes the planet to orbit the sun! Further, the direction of this rotation is opposite to that of all the other planets, except Uranus. As seen from the north celestial pole, all the other planets except Uranus orbit the sun and spin on their axes in counterclockwise directions. Venus' solar orbit is also counterclockwise, but its rotation is clockwise.

In October 1967, the space probes Venera 4 (U.S.S.R.) and Mariner V (U.S.A.) entered the atmosphere of Venus and transmitted their measurements back to earth. Their data were in general agreement that the planet's surface is very hot, probably above 500°F. They found that its atmosphere is composed predominantly of carbon dioxide (CO_2) and contains almost no oxygen or water. The atmosphere of Venus is very dense, and pressures recorded near the surface were about 20 times the atmospheric pressure on earth.

The fact that the atmosphere of Venus is primarily CO_2 makes it possible to explain its high surface temperature as being due to the "greenhouse effect." On earth we build greenhouses that are warmed above the temperature of the surrounding air by solar radiation. Light in the visible part of the spectrum enters a greenhouse easily through its glass windows, but the radiation absorbed and then re-emitted by the objects inside has wavelengths in the infrared region, which are not transmitted by glass, which is opaque to infrared rays. Thus the glass windows act as one-way valves to admit solar radiation but prevent its escape. The CO_2 atmosphere of Venus is transparent in the ultraviolet but absorbs strongly in the infrared; thus heat is trapped within the planet's atmosphere and its surface temperature has become very hot. We must conclude that not only is Venus too hot and dry to support life as we know it, but its extreme temperature and inhospitable atmosphere may make manned missions to its surface impossible. The "earth's twin" is very unlike the earth in a great many respects.

Mars

Mars is easily the most intriguing planet in our solar system. Visible to the unaided eye as a reddish object in the night sky, this namesake of the god of war has long been considered the most likely abode of any other intelligent life among the sun's planets. Unlike Venus, Mars has an atmosphere that is clear, and surface markings on Mars are easily seen through telescopes, though the interpretation of these markings has led to much controversy and speculation.

The diameter of Mars is 4180 miles, about half that of the earth. Mars takes 780 days to orbit the sun and rotates on its axis in 24 hours and 37 minutes. Its orbit about the sun is slightly more elliptical than that of the earth, and the distance of its closest approach to earth varies between 34,600,000 and 62,900,000 miles.

Two prominent white polar caps enlarge and shrink with the changing Martian seasons. Like the earth, Mars' axis of rotation is inclined with respect to its plane of rotation about the sun. This angle of inclination for Mars is about 25 degrees compared with the earth's 23.5 degrees; thus there are Martian seasons comparable to those we experience. As the seasons change, various areas on the Martian surface show changes of color, with reddish-brown spots giving way to blue-green patches.

Observations from earth of the surface details of Mars are extremely difficult, not because of problems in the Martian atmosphere, but because we must view the planet through the ocean of air that is *our* atmosphere. This atmosphere is turbulent and in constant motion, and any image that comes through it appears

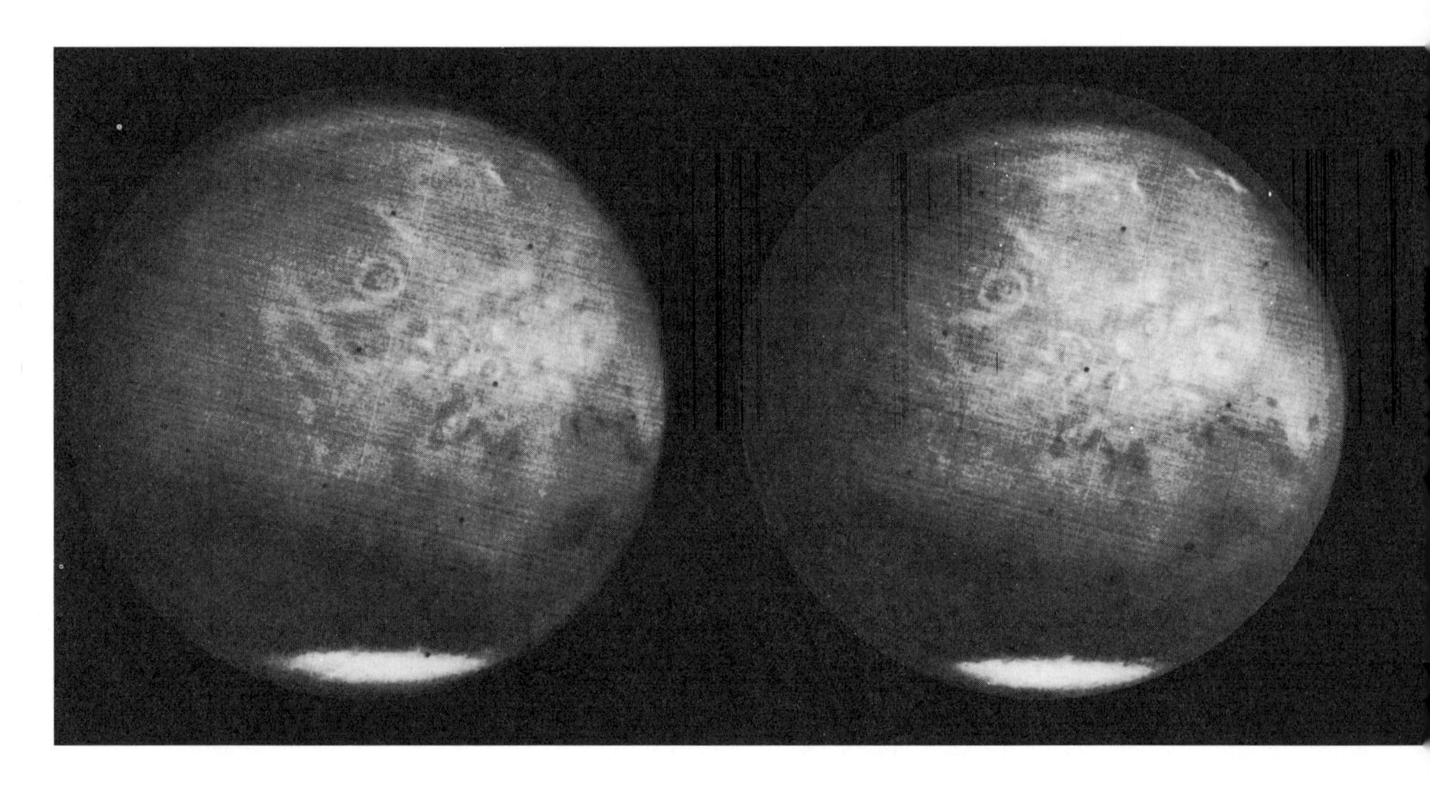

to dance in our view. (This is why the stars appear to twinkle.) Anyone who tries to photograph Mars through this blanket of air finds that an exposure for a period of time long enough to form an image on the film is also long enough to blur that image, due to the motion of our atmosphere. The visual observer, if he studies Mars over long periods of time, may find a few brief instants when the atmosphere between his telescope and the planet is relatively calm. During these moments of "good seeing," he may be rewarded by discovering what appear to be fine details in the surface markings of Mars.

The astronomer Schiaparelli reported that he had visually observed definite surface markings on Mars, which he described in his native Italian as *canali*, a word which should be translated as "channels" or "grooves." Though Schiaparelli did not consider these features to be of artificial origin, the American observer, Percival Lowell, called them *canals*, and suggested that they were the work of intelligent beings who had constructed them to carry water from the melting polar caps to the dry equatorial deserts. In his book, *The Canals of Mars,* published in 1907, Lowell concluded:

That Mars is inhabited by beings of some sort or other we may consider as certain as it is uncertain what those beings may be. . . . Apart from the general fact of intelligence implied by the geometric character of their con-

Two pictures of Mars, photographed by Mariner VII in August 1969. Note the southern polar ice cap. Pictures were taken 47 minutes apart, during which time Mars rotated 12 degrees. (Photographs courtesy Jet Propulsion Laboratory)

structions, is the evidence as to its degree afforded by the cosmopolitan extent of the action. Girdling their globe and stretching from pole to pole, the Martian canal system not only embraces their whole world, but is an organized entity.

In at least some quarters, speculation over the possibility of intelligent life on Mars reached a new high in 1967 with the proposal by the Russian astronomer Shklovskiy that Mars' inner satellite, Phobos, is a large hollow sphere and must be of artificial origin. His theory was based on observations which seem to indicate that the period of revolution of this satellite is slowly decreasing. If these observations are correct, Phobos is gradually moving into lower and lower orbits. The Martian atmosphere at the altitude at which Phobos orbits is far too thin to cause such an effect were the satellite solid and of the same approximate density as Mars.

This report also helped revive interest in the writings of Jonathan Swift concerning the Martian satellites. In *Gulliver's Travels,* published in 1725, Swift wrote of the discoveries of the scientists of Laputa, the island floating in the sky.

They have likewise discovered two lesser stars, or satellites *which revolve about* Mars; *whereof the innermost is distant from the Center of the primary Planet exactly three of his Diameters, and the outermost five; the former revolves in the Space of Ten Hours and the latter in Twenty-one and a Half. . . .*

It is interesting to note that these satellites were not actually seen until 1877. They are named Phobos (fear) and Deimos (terror), fit companions for the god of war. These satellites are actually located at distances of 2.3 and 6.7 diameters, respectively, from the planet's center. Phobos' period of revolution is 7 hours, 39 minutes, while the period of Deimos is 30 hours, 18 minutes. Phobos is the only satellite in our solar system that rotates about its planet in a time shorter than the period of revolution of the planet on its axis. An observer on Mars would see Phobos rise in the west and set in the east twice each day.

Swift's correct guess as to the number of satellites of Mars more than 150 years before they were discovered may have been based on a satellite-number scheme that had already been suggested in his day. Venus has no

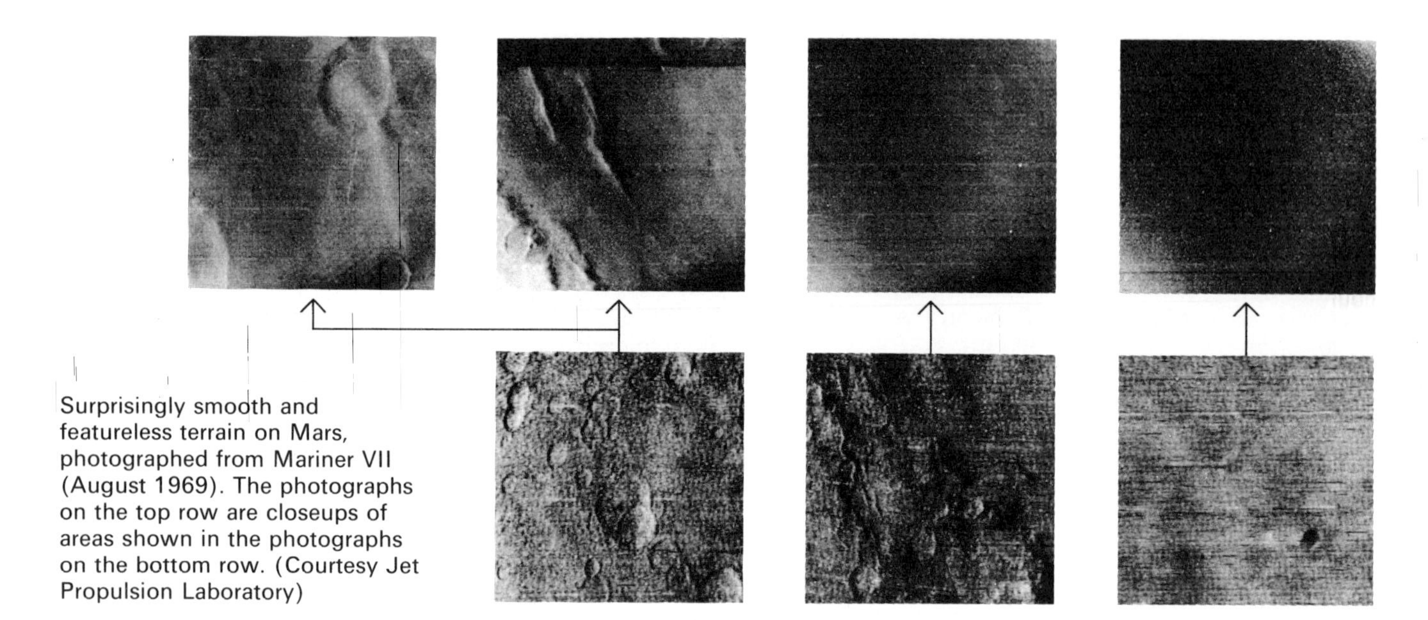

Surprisingly smooth and featureless terrain on Mars, photographed from Mariner VII (August 1969). The photographs on the top row are closeups of areas shown in the photographs on the bottom row. (Courtesy Jet Propulsion Laboratory)

satellites, the earth has one, and Jupiter—two places beyond Mars in the Bode's-rule scheme—was then thought to have only four satellites. In this progression it was logical that Mars should have two satellites. Even today we have not been able to precisely determine the size of the Martian satellites, though calculations based on the assumption that they reflect light in the same way as Mars itself have indicated that Phobos has a diameter of about ten miles, with the diameter of Deimos about half as great. The image of Phobos can be seen in several of the photographs taken by the space probe Mariner VII as Phobos passed between the cameras and the surface of Mars. A preliminary analysis of these pictures indicates that Phobos is not spherical, but elongated in the plane of its orbit, measuring roughly 14 by 11 miles. Phobos is larger than earlier telescopic measurements had indicated because it is apparently the least reflective object yet seen in the solar system. Why Phobos' surface should be so dark remains a mystery.

In July 1965, the U.S. space probe Mariner IV flew to within 6100 miles of the Martian surface and transmitted back to earth a series of pictures covering approximately 1% of the planet's surface. During late July and early August 1969, Mariners VI and VII passed within 2100 miles of Mars and sent back high-quality pictures of more than 10% of the surface. None of these pictures showed any canals, though three distinctly different types of surfaces were revealed. A part of the surface of Mars is heavily *cratered*, with some of the craters resembling those seen on the moon. Other craters, however, show important differ-

Top: Cratered region of Mars. Bottom: Chaotic terrain of Mars. This large-scale terrain feature is not found anywhere on either the earth or the moon. Both photographs were taken by Mariner VI in July 1969. (Courtesy Jet Propulsion Laboratory)

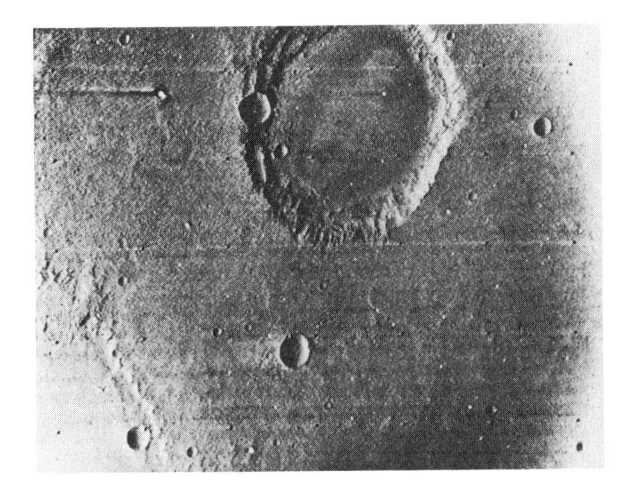

ences from lunar craters. Still another region of the surface of Mars is best described as *chaotic*. This terrain, consisting of regions of short, jumbled ridges and valleys, has no counterparts on either the earth or the moon. Perhaps the most puzzling topographical features seen are the large *smooth* surface areas. A region known as Hellas, which is about 1000 miles across, appears amazingly smooth and uncratered, though it is surrounded by cratered regions.

The presence of a Martian atmosphere is evident in the pictures taken by all three Mariner probes. Comparisons of the Mariner VI and VII photographs, taken five days apart, show clear evidence of atmospheric activity on Mars. Although this atmosphere was known to contain very little oxygen, a most surprising discovery was that it also contained little or no nitrogen. The atmosphere of Mars is composed almost entirely of carbon dioxide, with carbon monoxide being the second most abundant constituent.

The controversy over whether the polar caps of Mars are composed of snow and water ice or frozen carbon dioxide (dry ice) remains unresolved. Small amounts of water vapor have been detected in the atmosphere, but whether or not water is present in sufficient quantities to produce the polar caps is open to debate. Among the arguments against the hypothesis that the poles are covered with ordinary snow is the problem of how the tiny amount of water in the Martian atmosphere could be transported from pole to pole with the changing seasons. Since carbon dioxide is abundant in the atmosphere, no transport mechanism is necessary to explain dry-ice polar caps.

Mars is somewhat colder than the earth. Mariner IV measured the temperature of a Martian desert on a winter day to be −140°F. However, Mariner VI recorded a temperature of 73°F near the equator. As in the case of Venus, no magnetic field has been detected on Mars.

The surface temperatures of Mars will not provide any great hardship for manned exploration of the surface of the planet, particularly in the equatorial regions. Surface explorers of Mars will need to carry their own oxygen supplies, however, and must be shielded from the ultraviolet radiation from the sun, which is not absorbed by the Martian atmosphere. However, in every respect the Martian environment appears far less hostile to earth men than that of the moon.

The absence of atmospheric nitrogen on Mars apparently precludes the existence of earthlike forms there, although, on the basis of the evidence currently available, the possibility that there exist other life forms on Mars cannot be eliminated. The Mariner spacecraft, flying rapidly past the planet at distances no closer than 2000 miles, were simply in no position to observe life on its surface, if it were there.

The Asteroids

Our solar system contains an estimated 50,000 asteroids, most of which have orbits lying between 2.1 and 3.5 AU from the sun. We may adopt 2.8 AU as their rough average distance, in agreement with Bode's rule, though many of their orbits are highly elliptical. Although they are sometimes referred to as minor planets, the asteroids are rather small objects by planetary standards. Ceres, the largest, has a diameter of only 470 miles. Pallas is the second largest, with a diameter of 300 miles, followed by Vesta with 240 miles. The sizes of some of the smaller asteroids cannot be directly measured from earth, and must be calculated from the amount of sunlight they reflect.

The orbits of many asteroids are not confined to the space between Mars and Jupiter. Hidalgo goes nearly as far away from the sun as Saturn, while Icarus approaches the sun more closely than Mercury does. Some asteroid orbits can pass relatively near the earth. In 1937 Hermes came within half a million miles of our planet.

Escape velocity from the asteroids is very small, and none of them can be expected to support an atmosphere. However, their low escape velocities make asteroids ideal targets for manned landings and explorations. In contrast to the large expenditures of energy necessary to slow a spacecraft to enable it to land on the moon and then blast off again, a mission to an asteroid would involve only a little more energy than is required to leave the earth and establish the same solar orbit as the asteroid. The small size of an asteroid would also permit the exploration of a large fraction of its surface in a single mission.

Because they lack atmospheres, the surfaces of asteroids should have remained uneroded and unchanged from their primitive state. It is possible that more information about the early history of the solar system may be found on the asteroids than on either the moon or Mars. Geographos, which passed within 5.6 million miles of the earth in 1969, has been nominated as a good choice for a manned mission because its greatest distance from the sun is less than that of any other known asteroid.

Mercury

Mercury, the innermost planet, has a diameter of 3025 miles and orbits the sun in 88 days. Its orbit is highly elliptical and its distance from the sun at closest approach (its *perihelion* point) is 28,700,000 miles, while its greatest separation from the sun *(aphelion)* is 43,600,000 miles. The position of Mercury's perihelion point advances at the rate of 9 minutes and 34 seconds of arc per century, while the calculated change, due to the effects of all the other bodies in the solar system, is only 8 minutes, 51 seconds of arc per century. This discrepancy of 43 seconds of arc per century was one of the solar system's great mysteries until it was explained by Einstein's theory of relativity. (We shall discuss this problem in Chapter 3.)

It was long thought that Mercury required 88 days for rotation on its axis, and that it therefore always kept the same face toward the sun. Radar observations have now shown that the actual rotational period of the planet is two-thirds of its orbital period, or 59 days. This means that the Mercurian day is 175 earth days long, which leads us to expect that its surface temperatures are very hot during its long day and very cold during the night.

The velocity of escape from the surface of Mercury is 2.6 miles per second, only $1\frac{1}{2}$ times the escape velocity from the moon. Its low escape velocity combined with its high temperatures would normally indicate that Mercury has long since lost any atmosphere it once possessed. However, temperatures measured on the dark side of the planet are much higher than expected. One way to explain this fact is to assume that there is an atmosphere present to aid heat transfer from the sunlight to the dark side. It has been suggested that the atmosphere of Mercury may be constantly replenished by hydrogen expelled from the sun.

Recent radar studies have indicated the presence of several large topographic features on the planet's surface. These features appear to be almost the size of continents, and fixed on the surface. Similar rough surface features have been seen on Venus, but those on Mercury appear larger, relative to the size of the planet.

The large inclination of Mercury's orbit to the ecliptic and its proximity to the sun make flights to the innermost planet quite demanding from an energy standpoint. Missions involving swing-bys of Venus will somewhat lessen the energy requirements, though they will require more time to complete than direct flights to Mercury. Spacecraft capable of reaching Mercury must have special thermal shielding to operate that near the sun. However, from what is now known of Mercury, manned landings on it, near the zone of transition from day to night, may be possible.

The Jovian Planets

Mercury, Venus, the earth, and Mars are often referred to as the *terrestrial planets* because of their similarities in size, density, and other characteristics. Jupiter, Saturn, Uranus, and Neptune also exhibit a marked family resemblance, and may be called the *Jovian planets*, after Jupiter, the dominant planet of the solar system. The terrestrial planets have average densities between four and five and one-half times that of water, and all have diameters between 3000 and 8000 miles. The average density of the Jovians is only slightly greater than that of water, while their diameters range from 27,600 miles (Neptune) to 88,700 miles for the giant Jupiter. Despite their size, the Jovians rotate rapidly on their axes. A day on Jupiter lasts less than 10 hours, and the longest of their periods of rotation is that of Neptune, at about 16 hours.

The Jovians have deep atmospheres, with methane, ammonia, and hydrogen gases abundant. Their surface temperatures are all apparently below −220°F. Jupiter has been found to radiate energy at radio frequencies. It is particularly active at wavelengths between 10 and 30 meters, and anyone with a radio tunable to these bands may listen to bursts of static-like noise from Jupiter.

The Jovians all possess multiple satellites: Jupiter has 12, Saturn 9, Uranus 5, and Neptune 2. Jupiter's four largest satellites are approximately the size of our moon, as is Saturn's sixth satellite, Titan, and Neptune's Triton. Titan may be unique in the solar system, for it is the only satellite known to have an atmosphere.

The problems associated with manned exploration of the surfaces of the larger planets, particularly Jupiter, appear formidable. A human explorer on Jupiter would face extremely low temperatures, high pressures in a lethal atmosphere and a gravitational attraction two and one-half times that of earth. The first close-up information from Jupiter should be obtained from an unmanned probe now scheduled for launch in 1972. Among the fascinating questions we hope will be answered is whether or not Jupiter is radiating more energy than it receives from the sun.

Pluto

Pluto's great distance from earth ensures its role as the most mysterious of the planets. Thus far, attempts to measure even such basic data as the size and mass of the planet have only led to paradoxical results. The apparent diameter of Pluto, based on visual observations, is about 3600 miles. Yet calculations of its mass, derived from measurements of its effect on the orbits of Neptune and Uranus, lead to the conclusion that Pluto is about three times as dense as gold! (A given volume of gold weighs almost twice as much as an identical volume of lead.) Pluto, like Mercury and Venus, has no known satellites. And unless Pluto has an appreciable amount of internal heating, it appears too bitterly cold to support any form of life as we know it, for sunlight there is only 1/1600th as brilliant as on earth.

2.6 WHY GO INTO SPACE?

Reactions to man's ventures into space thus far have ranged between the extremes of those who suggest that space flights should never have been attempted to those who insist that we have as yet done far too little. Unfortunately, the logic associated with much of this reasoning has been of the sort that either asserts, "God never intended for man to leave this earth" or that which says we should go into space simply "because it is there."

It is a mistake to lump all that which has been done and can be done in space into the same category. There are at least four different important facets to any space program: (1) military, (2) utilitarian, (3) pure science, and (4) adventure.

Military Applications

On October 4, 1957, the U.S.S.R. launched into orbit the earth's first artificial satellite, Sputnik 1. At that time, the outgoing Secretary of Defense, Charles E. Wilson, sought to reassure the American public by insisting that "nobody is going to drop any bombs on you from a satellite." While nobody has yet "dropped" anything from a satellite, to be perfectly candid, the military implications of space have given both the U.S. and the U.S.S.R. space programs their greatest impetus. This is true in our own country despite the fact that the primary space agency, the National Aeronautics and Space Administration (NASA), is ostensibly a civilian agency.

One of the primary reasons for the military interest in satellites is their usefulness in reconnaissance and surveillance. An amazing amount of detail can be resolved from orbital altitudes of 100 miles and upward. One scientist working in the field of military satellite reconnaissance has boasted that, given perfect weather conditions, U.S. "spy-in-the sky" satellites can read the rank on a soldier's uniform from an altitude of 300 miles.

Photographs taken from space and relayed back to earth have revealed missile sites and other equipment deployment. It is possible to equip satellites with infrared sensors that will detect rocket launchings from the earth's surface. It may be possible to use satellite stations as tracking and guidance links in ABM systems, and missiles may even be launched from satellites in orbit. No small part of the interest in using satellite surveillance and guidance systems is the fact that they are virtually invulnerable to ground-based oppo-

The mid-East: Egypt, the Sinai peninsula, Syria, Arabia, photographed from Gemini XI (September 1966). Smoke from an oil fire (and its shadow) are easily seen at this altitude (triangular dark area at upper right). (Photograph courtesy of NASA)

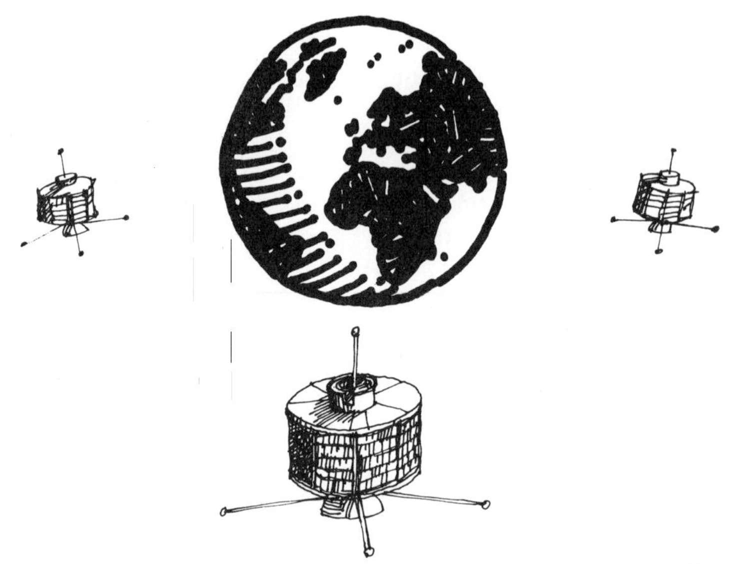

FIG. 2.9 Synchronous satellite communications network.

nents. Sending a missile into space to destroy an existing satellite demands technical capabilities which the U.S. has not yet demonstrated.

Finally, satellite relay facilities make possible global military communications networks that will not contain vulnerable ground links. With the use of highly directional antennas—or even better, laser beams—for transmission and reception, interception of messages would be all but impossible.

Utilitarian Applications of Space

One of the most beneficial applications of satellites to date has been in the area of communications. *Synchronous satellites* (Fig. 2.9), in place above the Atlantic, Pacific, and Indian Oceans, are now being used to provide around-the-globe communications networks, and live color TV broadcasts between the United States and Europe have become almost commonplace. In July 1969, transmission to the U.S. of television coverage of the investiture of Prince Charles of England as Prince of Wales was threatened due to problems with one of the Atlantic Ocean communications satellites. The program was therefore rerouted via the Indian and Pacific Ocean satellites, proving that television transmission around the world is entirely feasible.

A proposal has been made to locate synchronous satellites above the U.S. for handling

Hurricane Gladys, about 150 miles southwest of Tampa, Florida; photographed from Apollo 7. (Photograph courtesy of NASA)

domestic telephone and TV transmission during the 1970's and beyond. Not only would this proposed system greatly increase the capacity of our communications networks, but it has been calculated that these satellite systems would be cheaper than equivalent ground-linked systems.

Another use of satellites is observed almost daily in the weather pictures that are transmitted back to earth and often seen on TV weather-news telecasts. Before the first weather satellite was put into place, less than 20% of the earth's surface was covered by regularly reporting weather stations. Satellite pictures can now cover the entire surface of the earth in a matter of hours. The first weather satellite, TIROS III, had been aloft only a short time in 1961 when it discovered Hurricane Esther, several days earlier than this storm would have been detected by conventional methods. These satellite photos have made possible the development of far more accurate long-range weather forecasting methods than would have been possible otherwise, and have also increased our ability to predict severe weather patterns.

The Grand Bahama Bank, photographed from Gemini V (August 1965). The clarity of the ocean-bottom topography suggests the use of space photography as an aid in oceanographic investigations. (Photograph courtesy of NASA)

Map-making is tremendously speeded up by satellite photos. In over 40 years, less than one-fourth of Peru had been photographed from the air, yet in just three minutes the astronauts of Gemini IV were able to make high-resolution photographs of three-fourths of that country. Measurements of quantities such as snowfall coverage—in order to predict runoff and subsequent down-river flooding—are much more easily and safely done by satellites than by low-flying airplanes.

Many other uses of satellites may be made in oceanographic, geologic, air and water pollution, and even area-planning studies. Infrared sensors and multicolor photographs from space can pick out regions of blighted trees in forests, determine soil temperatures so that agricultural engineers will be able to advise farmers as to the best times to plant crops, help those in the fishing industry to locate schools of fish, and indicate the presence of underground mineral deposits. A series of Earth Resources Observation Satellites (EROS) is expected to be flying by 1974.

The manned-space-flight program has also led to the development of many products and devices that have had widespread application in more utilitarian areas. Among these products are biomedical probes and sensors, insulation and heat-shielding materials, and instrumentation for many specialized tasks.

Scientific Experimentation

Although many projects which began as matters of purely scientific investigation have been applied in very useful ways, the researcher is continually interested in those things which may be learned from space data that could have been found in no other way. From the time the very first artificial satellites were put into orbit, we were able to learn much about the exact shape of the earth. We knew that it was far from being a perfect sphere and that it was rather flattened at the poles, but we have now been able to measure the bumps and bulges and surface irregularities of our globe. Distances between points on earth may now be measured with far greater accuracy

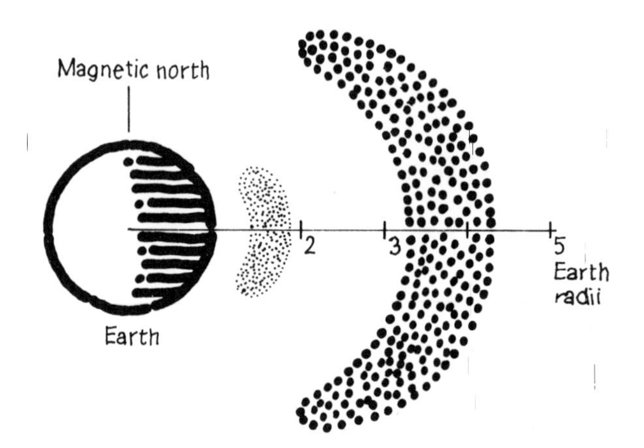

FIG. 2.10 Van Allen radiation belts.

than ever before, and questions about the motion of the continents across the face of the globe may now be answered. From satellite-borne detectors, we found that there are electrically charged particles, trapped by the earth's magnetic field, orbiting the earth at altitudes of about 500 miles. These particles make up the Van Allen radiation belts (Fig. 2.10). Space probes have also gone out from our planet to discover the Solar Wind—composed of effluxes of gas and particles ejected by the sun—that is rushing past the earth.

Because our atmosphere absorbs radiation in the ultraviolet and infrared wavelengths, it is impossible for us to observe signals at these wavelengths (which reach us from space) when we are on the earth's surface. Astronomers are now analyzing satellite data taken from an Orbiting Astronomical Observatory (OAO). Operating above any atmosphere that might distort its view, a moderate-sized telescope in space can show more detail of our universe than the most powerful ones on the earth's surface can reveal.

In laboratories in space, experiments may be performed and measurements made that cannot be done on earth. Space provides an enormous vacuum chamber in which both very high and very low temperatures are available in abundance. Moreover, such things as true three-dimensional casting of materials can be performed only under conditions of zero gravity.

The unmanned space probes and, later, manned missions that will be sent to the other bodies in our solar system will reveal information that may provide answers to many long-unanswered questions. What are the other planets like? Is there any form of life on any of them? From studies of the other planets and their satellites we may learn much about the history and makeup of our own planet. A study of any extraterrestrial life forms found may be of the utmost significance in helping us understand more about our own life processes.

Man into Space: The Greatest Adventure

There are many reasons for sending men into space. For, although certain processes may be automated and done by machines, no machine has yet been made that is so adaptable and versatile as man. An automated device can function properly only when it has been programmed to ask the right questions and look for the right things, but a trained human observer may be able to obtain useful information when he encounters conditions very different from those expected. For example, a machine might pick up *any* lunar rock and return it to earth for analysis, but only a man can choose the most interesting rocks or determine whether he has selected a representative sample.

Still, using men to explore space may be a luxury unjustifiable in view of the severe penalties that must be paid for providing human-life-support systems in space vehicles. Manned spacecraft must be significantly larger, heavier, and much more costly than unmanned probes, and the differences between the two types of craft increase sharply as the mission is lengthened.

There is yet another reason for venturing into space. Space exploration offers adventure of the highest order. The landing of men on the face of the moon and their safe return to earth represented only a small, timid venture into the vastness of space, but it was truly a "giant leap for mankind" in demonstrating that men can travel from this planet and live and function on another body in space. Manned exploration of the planets represents possibly the most difficult and costly physical project ever attempted, and this alone would be sufficient reason for the adventuresome to seek to loose the bonds of this earth and soar among the planets.

It is pointless to speculate over any benefits that may accrue from discoveries yet to be made in space—including even so great a finding as the presence of life elsewhere in our universe—for discoveries cannot be predicted and their existence or non-existence after the fact cannot be used to justify mission planning. Space exploration is adventure of the highest order, and many wish to participate in it.

FOR MORE INFORMATION

Fred L. Whipple, *Earth, Moon and Planets,* Cambridge, Mass.: Harvard University Press, 1968. A highly readable survey of the solar system.

Gerald Holton and Duane H. D. Roller, *Foundations of Modern Physical Science,* Reading, Mass.: Addison-Wesley, 1958; Part III: The Study of Planetary Systems. An historical and philosophical look into planetary motion and the structure of the solar system.

Samuel Glasstone, *Sourcebook on the Space Sciences,* Princeton, N.J.: D. Van Nostrand, 1965. An encyclopedic volume that will probably tell you more than you want to know about almost anything connected with space. Some of the facts included are now outdated, however.

Gerard P. Kuiper and Barbara M. Middlehurst (editors), *Planets and Satellites,* Chicago, Ill.: University of Chicago Press, 1961. A scholarly review of studies on the solar system.

George E. Wukelic (editor), *Handbook of Soviet Space-Science Research,* New York: Gordon and Breach Science Publishers, 1968. A look at Russian exploits in space.

Percival Lowell, *Mars and Its Canals,* New York: MacMillan, 1907 (dedicated to G. V. Schiaparelli). Lowell's arguments for life on Mars.

Jonathan Swift, *Gulliver's Travels,* New York: Heritage Press, 1940. Chapter 3 of Part III tells of the Laputian astronomical observations.

A great many booklets are published by NASA, summarizing up-to-the-minute data from space. These booklets are available at modest cost from the U.S. Government Printing Office in Washington, D.C.

QUESTIONS

1. Why do planets appear to move with respect to the stars in the sky?
2. Which of the planets have been known since antiquity? How recently was the last of the planets discovered?
3. According to Bode's rule, where would you expect the position of the tenth planet to be?
4. Explain why Pluto is sometimes not considered a legitimate planet.
5. Can an air-breathing engine operate as a reaction propulsion system? Explain.
6. What is the essential difference between launching a space vehicle from earth to reach the inner planets (Mercury, Venus) and one that is to travel to the outer planets (Mars, Jupiter, etc.)?
7. Describe the proposed Grand Tour of the Planets.
8. Which of the planets moves in the most elliptical orbit about the sun? Which planet has the most nearly circular orbit?
9. On the basis of surface-level escape velocity *alone*, which of the planets would be least expected to have a significant atmosphere? Which should have the most atmosphere?
10. Give reasons for describing the moon as a minor planet rather than just the earth's satellite.
11. A planet named Vulcan was once thought to exist inside the orbit of Mercury. Why would this planet have been extremely difficult to observe?
12. If Vulcan did exist, what physical characteristics would you expect it to have?
13. Review the "evidence" for the existence of intelligent life on Mars.
14. How far wrong was Swift in his description of the orbits of the moons of Mars?
15. It has been suggested that the first manned expedition to Mars should land on Phobos or Deimos rather than on the planet itself. What reasons can you give to support this suggestion?
16. The asteroid Ceres orbits the sun at 2.77 AU. When it was first discovered, it was hailed as the missing planet in the Mars-Jupiter gap. Ceres is proportionately larger, in comparison with Mercury, than Mercury is in comparison with Jupiter. Why is Ceres not considered a planet?
17. What is particularly attractive about the asteroid Geographos as a potential observation post in space?
18. Pluto is sometimes classified as being among the terrestrial planets. What points of similarity can you find between Pluto and earth? What differences?
19. Describe some potential benefits of the EROS satellites.
20. What have manned missions to the moon accomplished thus far that could not have been accomplished with unmanned probes?
21. Which of the objects in the solar system can you expect to be the subject of manned investigations during your lifetime? Why?

PROBLEMS

1. Astronauts who have made "space walks" have used small propulsion units for maneuvering in zero gravity. If a spaceman were to lose his propulsion unit, is there any way he could propel himself through zero-gravity space?

2. Our spacecraft have flown by Mars without seeing signs of intelligent life on the planet. If a

Martian spacecraft had flown by earth at about the same altitude, what signs of life on our planet do you suppose might be seen?

3. Imagine that you are exploring the moon's surface when the Lunar Land Rover you are driving breaks down. Your spaceship is now over the lunar horizon and you will have to walk back to it. You have a number of items with you which you may find useful on your walk back. Examine the following list of articles and decide which you would take with you and which you would leave and give reasons for your choices. Items: extra oxygen tanks, a box of matches, flare pistol and flares, large umbrellas, a magnetic compass, a bullhorn, a star-chart, a clock, an alcohol stove, a flashlight, a solar-powered heater.

4. To help astronauts pass their time during future long voyages in space, the suggestion has been made that they participate in games and sports aboard their space vehicles. Some familiar games might take on an entirely new character in space, as, for example, three-dimensional billiards or six-wall handball. Think of either modifications of familiar games or entirely new ones that would be particularly adaptable to conditions of zero gravity.

5. Assuming that the funds appropriated by Congress for space programs would not be appropriated for other purposes if there were no space program, list the space programs you feel should be supported during the remainder of this century, and defend your choices.

6. It has been suggested that one of the early Mars expeditions attach a large engine to one of the Martian satellites, bring it back and put it into earth orbit. There it could be thoroughly studied much more easily and less expensively than if an equivalent number of trips were made to Mars. Attack or defend this proposal.

7. The Martian moon, Deimos, is only about five miles in diameter. Let us say that, in the late 1980's, a NASA expedition captures Deimos and places it in an orbit near earth, where it can be the object of intense study. In 1990 it is declared government surplus property, and shortly thereafter Disney Universe announces its acquisition and reveals plans for a space-based tourist attraction. Visitors are to be shuttled up and back by rocket to visit this out-of-this-world tourist and convention center. Suggest some of the difficulties that would be encountered by the developers.

8. Assume that you are the leader of an expedition from outside the solar system. It is your responsibility to survey the earth and plan manned landings on its surface. Would you choose to land on the blue areas or on the brown areas? Why? (While all U.S. manned flights to date have landed in the oceans, all the Russian flights have come down on land.)

9. No fewer than six nations now own satellites that are in earth orbit, in addition to the privately owned communications satellites. It has been assumed by some that the laws of the high seas should also apply in space. Among the features of these laws is the right to salvage abandoned vessels and proportionate responsibility in damage claims. Is this a reasonable suggestion?

10. In view of the discussion about the limits of manned exploration in the universe, what can you conclude about the theory that UFO's have extra-terrestrial or even extra-solar-system origins?

11. The first soil samples returned to earth from the moon contained large amounts of titanium ore. The Board of Amalgamated Bullfinch, Inc., of which you are an executive, has just decided to explore the possibilities of mining moon-titanium. Discuss the feasibility of this project in terms of what you now know, and list those questions you want answered before you make a recommendation on this project.

12. A fractional orbital rocket system could reduce the time required to travel from New York to Los Angeles to about 20 minutes. In view of the fact that supersonic transport (SST) aircraft may soon be in commercial service, discuss a passenger-carrying rocket system as the next step in high-speed, long-range transportation.

CHAPTER 3 SPACE-TIME AND OUR UNIVERSE

3.1 TOWARD HIGHER SYMMETRIES

One measure of our progress in understanding nature is the degree to which we are able to deal with higher and higher symmetries. So long as we have to consider each and every observed physical phenomenon as a separate case, it is obvious that our picture of the universe is very incomplete. It is only when we are able to group phenomena together, to say that the consequences of a certain physical law are observed in a given set of phenomena, that we have begun to understand what really happens.

It was a signal triumph when Isaac Newton was able to understand that just as an apple falls from a tree toward the earth, the moon in its orbit is also falling toward the earth and both are examples of gravitational interactions. Before Newton's time, our world appeared to have two-dimensional spatial symmetry in the horizontal plane, with some different laws governing vertical motion. The fact that things fall down distinguished the up–down direction from the two horizontal ones. It was an understanding of the gravitational interaction that removed this apparent difference, for we now realize that the three spatial dimensions in our world are quite symmetric, and that things fall down simply because the direction in which the gravitational force acts between an object and the earth is that direction we define as "down."

But even in a world of complete spatial symmetry, the concepts of space and time were once considered quite distinct from one another. It was long thought that time could be measured in a completely satisfactory fashion by some great universal clock. To uniquely locate any object it is necessary to specify both its spatial and time coordinates, but it was felt that a single "standard" clock would suffice for all time measurements. This view had to be changed. A still-higher dimensional symmetry was introduced through the genius of Albert Einstein.

3.2 THE RESTRICTED THEORY OF RELATIVITY

During the year 1905 Albert Einstein worked in a relatively obscure position as a clerk in the Swiss Patent Office. Yet in that single year he published three monumental papers that established him for all time as a physicist of the first rank. One of those papers concerned

Albert Einstein (1879–1955). (Photograph by Fred Stein)

the Restricted Theory of Relativity. From this work came a strikingly changed conception of the universe in which we live.

The "restriction" referred to in the name of the Restricted Theory of Relativity is that only systems and observers moving with constant velocities with respect to each other are considered in this theory. If a physical situation involves an acceleration, the General Theory of Relativity must be invoked.

The Restricted Theory of Relativity deals with the problem of comparing measurements made in different *inertial systems* by observers who are moving at a constant speed with respect to each other. To understand what we mean by an inertial system, let us consider that you are seated in a room, stationary with respect to the earth, as you read this page. For your purposes the room in which you are located is your inertial system and you could carry out all sorts of laboratory experiments in it if you had the equipment and inclination.

Now suppose that there is at this moment an airliner flying overhead at 500 miles per hour.

With respect to the earth, the airliner is obviously in motion. Yet, if you were seated inside

FIG. 3.1

the aircraft on a day when the air was very smooth and the plane was flying at a constant speed, you might not be aware of this motion. Under these conditions the inside of the airplane could be as convenient an inertial platform as that "stationary" room you were in, back on earth. Actually, the room on earth is not so stationary as one might suppose, for the earth is spinning on its axis rapidly enough that a point on its surface at the equator is moving at about 1100 miles per hour. The whole earth travels at about 6600 miles per hour in its orbit about the sun, and the sun is itself moving through our galaxy at 43,000 miles per hour.

All the conclusions of the Restricted Theory (often called the Special Theory) of Relativity follow from two postulates:

1. *There is no preferred inertial system.*
2. *The speed of light in free space has the same value in all inertial systems.*

The speed of light in free space, normally denoted simply by *c*, is about 186,000 miles per second.

The first postulate tells us that since there is no preferred inertial system, any inertial system is as good as any other for making measurements. We should be thankful that this is the case, for otherwise a state of utter chaos would exist, with the results of physical experiments depending on the lab in which they were done. This democracy among inertial systems may be understood in connection with the fact that all observers measure the same value for the speed of light. To illustrate what we mean, let's first think about a situation that does not involve relativity.

Suppose that there are two boats on a lake, caught in a fog so dense that no surrounding landmarks may be seen (Fig. 3.1). Further suppose that one of the boats is stationary in the water, while the other is moving at a small, constant speed as the observers in the two boats pass one another. As the boats pass, it is obvious that they are moving relative to one another, but is there any way of telling *which* of the boats is moving? (Or are they both moving?) There is a very simple way to answer both these questions. Just as the boats drift alongside each other, let one of the observers drop a rock or some other object into the water. Waves will go out in all directions in concentric circles from the point at which the rock splashed into the water, the front of each wave moving at a constant speed.

To determine which of the boats is moving, it is only necessary to examine the pattern of the circular waves around each boat. The boat that is stationary will be at the center of the circular wave pattern and will remain there as time passes. But if the boat is moving, after a few seconds any given wavefront will be nearer one end of the boat than the other. By means of these observations, the two observers may uniquely determine their absolute motion through the water.

This type of measurement cannot be made using light, however, since each observer must measure the same constant value for the speed of light in free space. As an example of this fact, suppose that the preceding experiment is repeated, but with a flare being set off as the boats pass, instead of a rock being dropped into the water (Fig. 3.2). Light

FIG. 3.2

goes out from the flare in a constantly expanding sphere, but whether his boat is stationary or moving, an observer will see the light waves go out in special patterns, with his boat located at the center of these spheres. This occurs because each observer must measure the same value for the speed of light regardless of his motion. Thus all the light emitted by the flare must recede from the observer at the same constant speed.

Now the above situation may lead to apparent logical inconsistencies. One might ask what would happen if an object were approaching an observer at half the speed of light, and then turned on the "headlights." Why wouldn't the observer measure the speed of light to be the sum of the vehicle's speed plus the speed of light? That is, wouldn't the beam approach him at 1.5 times the speed of light?

The answer to our question lies in the fact that there is a fundamental difference between light waves and water waves. Water waves involve motion of the water and the water is necessary as a supporting medium for the propagation of these waves. Light, however, can move through totally empty space and does not require a supporting medium.

This is not an easy concept to accept, for, by analogy with water waves and sound waves, it seems only reasonable that light waves must be propagated through *something*. Until late in the nineteenth century, scientists supposed that this "something" was an odorless, colorless, tasteless—absolutely undetectable—fluid that filled all space. This fluid, which they called *ether*, was the medium through which they believed that light was propagated.

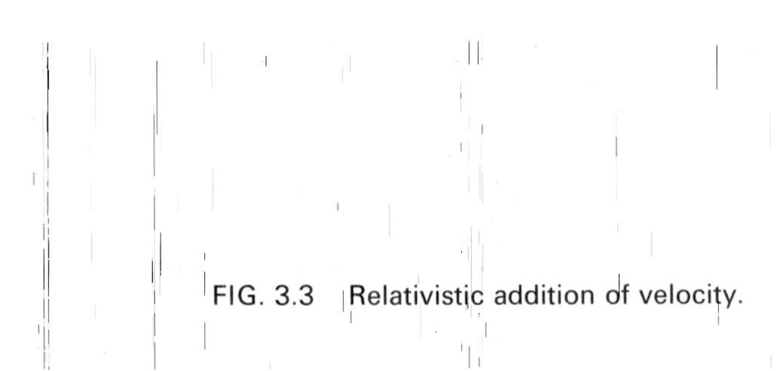

FIG. 3.3 Relativistic addition of velocity.

The downfall of the ether theory was the result of a series of experiments performed in 1887 by the American physicists, A. A. Michelson and E. W. Morley. They reasoned that, if all space is filled with ether, they should be able to detect the motion of the earth through that ether. But the inescapable conclusion of their work was that light is propagated through completely empty space. There is no ether!

When we realize that this is the situation, there is no *a priori* reason why we should *expect* light to behave like water waves or sound waves, and numerous experiments confirm that it does not. Thus an observer always measures the same value for the speed of light

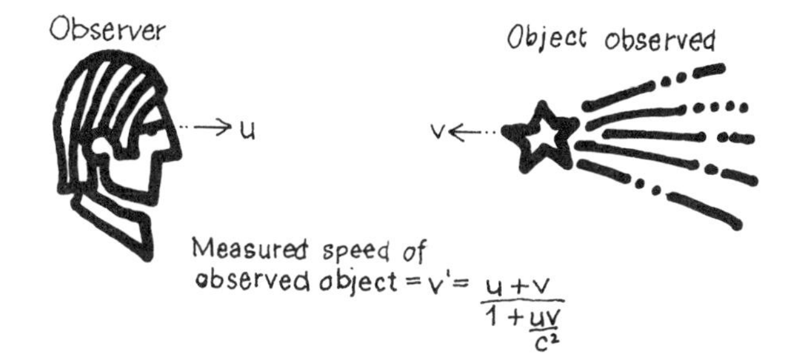

If $u = 0.5c$ and $v = 0.5c$,

$$v = \frac{0.5c + 0.5c}{1 + \frac{(0.5c)(0.5c)}{c^2}} = \frac{1.0c}{1 + 0.25\frac{c^2}{c^2}} = \frac{1.0}{1.25}c = \boxed{0.8c}$$

If $u = 0.5c$ and $v = c$,

$$v' = \frac{0.5c + c}{1 + \frac{(0.5c)(c)}{c^2}} = \frac{1.5c}{1 + 0.5\frac{c^2}{c^2}} = \frac{1.5}{1.5}c = \boxed{c}$$

in free space, *c*, no matter what the relative velocities of the light source and the observer happen to be.

The equation shown in Fig. 3.3 may be used to correctly calculate the addition of two relativistic velocities. This equation, which is known as the *relativistic velocity addition law,* may be used to find the relative velocity one would measure for any two objects, whether or not one is moving at the speed of light. Note that this equation correctly predicts that the value *c* will be obtained for the speed of light in free space, regardless of the velocity of its source.

3.3 RELATIVISTIC MEASUREMENTS

Let us now deal with the problem of two observers in laboratories in different inertial frames who may carry out whatever experiments they think proper. The question we wish to answer is: How will the results obtained by these two observers compare?

Surely if physical laws are valid at all, they must not depend upon the location at which an event occurs. Otherwise there would be no point in writing down physical laws at all, since they would all be of only local applicability. Fortunately, the forms of all the physical laws investigated by the two observers will be found to agree *within their own inertial systems.* Thus if one observer carries out an experiment in his inertial system, his colleague may repeat the experiment in a different inertial system and obtain the same results. This is what is meant by the statement of the first postulate of the Restricted Theory that "there is no preferred inertial system."

However, when our observers who investigate phenomena in different inertial systems try to compare their results, confusion arises. One might think that both observers could at least use the same time standard, such as Greenwich Mean Time, with only the addition or subtraction of a fixed number of hours to agree with local sun-time zones. Yet the fact of the matter is that the two observers in different inertial systems can't agree on *anything:* not on the *length* of a ruler, or the *speed* of an object, or the *time* or even the *order* in which events take place—unless they consider the laws of relativity and make appropriate adjustments to the values that they observe.

This most perplexing state of affairs stems from the fact that no information may be transmitted from one system to the other faster than the speed of light, and this speed of light is infinite. To be sure, the speed of light is very great, and the disagreements in the measurements we shall be describing will be so very small that unless the objects involved are traveling at speeds near that of light they may normally be ignored. Yet all these effects are real and have been verified many times in laboratory experiments.

Length Contraction

Let us first try to compare the measurements of length of two objects. Suppose that our two observers—we'll call them *A* and *B*—and their inertial systems are moving at constant speeds with respect to each other, so that each is able to see what the other is doing. Now each observer will naturally feel that he is the one at rest and that the other fellow is doing the moving; this is precisely where the idea of relativity begins to come in. Of course if one observer were in the airplane we mentioned a moment ago and it were to land, it would be obvious to him that he had been moving with respect to the other observer. But we can imagine that our observers are in space vehicles, far away from anything else in the universe, moving at a constant speed with respect to one another.

Let observers *A* and *B* each now perform a series of measurements using meter sticks that they have carried with them. Each finds that his meter stick is fine for making measurements in his own system, but when he tries to measure objects in the other system, something surprising happens. If *A* tries to measure the length of *B*'s meter stick, he says it is too short. The exact length that *A* measures for *B*'s meter stick depends on the relative speed of the two observers. If this speed is 0.6*c* (60% of the speed of light), *A* measures *B*'s meter stick to be only 80 cm long. However, *B* thinks that *his* meter stick is in good shape—100 cm long, just as it should be—but that *A*'s meter stick is only 80 cm long.

Now you may begin to see what we mean by relativity. Each observer thinks that his sytem is the one in which all measurements are correct and that there is something wrong with the other system. To understand the situation, you must view it in a relativistic way: The answers obtained in one inertial system are just as valid as those obtained in any other inertial system *so long as one stays in his own system.* It is when an observer in one system attempts to make measurements in a different system that he must apply *relativistic corrections* in order to meaningfully compare the values obtained.

An observer who wishes to have his measurements agree with those of the observer in the moving system simply multiplies his results by a correction factor. The size of this correction factor depends on the relative speed of the observers. Table 3.1 contains some of the values an observer would measure for the length of a meter stick in a system moving with respect to his own. Note that relativistic corrections are unimportant so long as the speeds involved are small fractions of the speed of light.

TABLE 3.1

Speed of system moving relative to observer	Length of meter stick in moving system, measured by stationary observer
0	100 cm
$0.01c$	99.995 cm
$0.1c$	99.5 cm
$0.5c$	87.1 cm
$0.6c$	80.0 cm
$0.8c$	60.0 cm
$0.9c$	43.5 cm
$0.99c$	14.1 cm

Time Dilation

Our friends, *A* and *B*, might also find it interesting to compare the *time* measurements they make. If *A* looks at *B*'s watch through a telescope, he may say, "*B*'s watch is slow." But *B* considers that his watch is in good order, although if he then looks through his telescope at *A*, he will say that *A*'s watch is the one that is slow.

If we are willing to consider time in the same way as we considered positional coordinates, this state of affairs is no more difficult to understand than the contraction of lengths. This is where the still higher symmetry in our view of the universe comes in: Relativistically speaking, we cannot simply locate an object by three spatial dimensions, but must locate it in *space-time*. Time then becomes a fourth dimension, as valid and as important as the three spatial dimensions. There is no longer a unique, universal time. The time each observer measures in his own inertial system is known as his *proper time*.

A striking experimental verification of both the length-contraction and time-dilation effects is seen in the behavior of a subatomic particle, the *μ-meson* (usually known as the *muon* for short). The muon may be formed in our laboratories in reactions initiated by very energetic particles coming from particle accelerators (see Chapter 7). Muons are not stable particles; they decay with a *half-life* of only 2 millionths of a second. This means that in a given sample of muons, half would be expected to decay in the first 2 millionths of a second, half of those that remained in the next 2 millionths of a second, and so on.

Muons are also created high above the earth by the energetic particles of cosmic radiation. A typical muon created in a cosmic-ray shower is known to travel at a speed of 2.994×10^8 meters per second, which is 0.998 of the speed of light in free space. Traveling at this speed, a muon could go only about 600 meters before its 2 millionths of a second half-life time would run out. But these muons are known to be created at altitudes of about 9500 meters above the earth's surface and many of these same muons are detected near the earth's surface. Since we know how fast they are going, we may easily calculate how long the muons live in traveling this distance. The result is that 31.7 millionths of a second are required to go 9500 meters at a speed of $0.998c$. This is almost 16 times as long as a muon should live!

This seemingly impossible state of affairs can be explained simply when one considers the relativistic length-contraction and time-

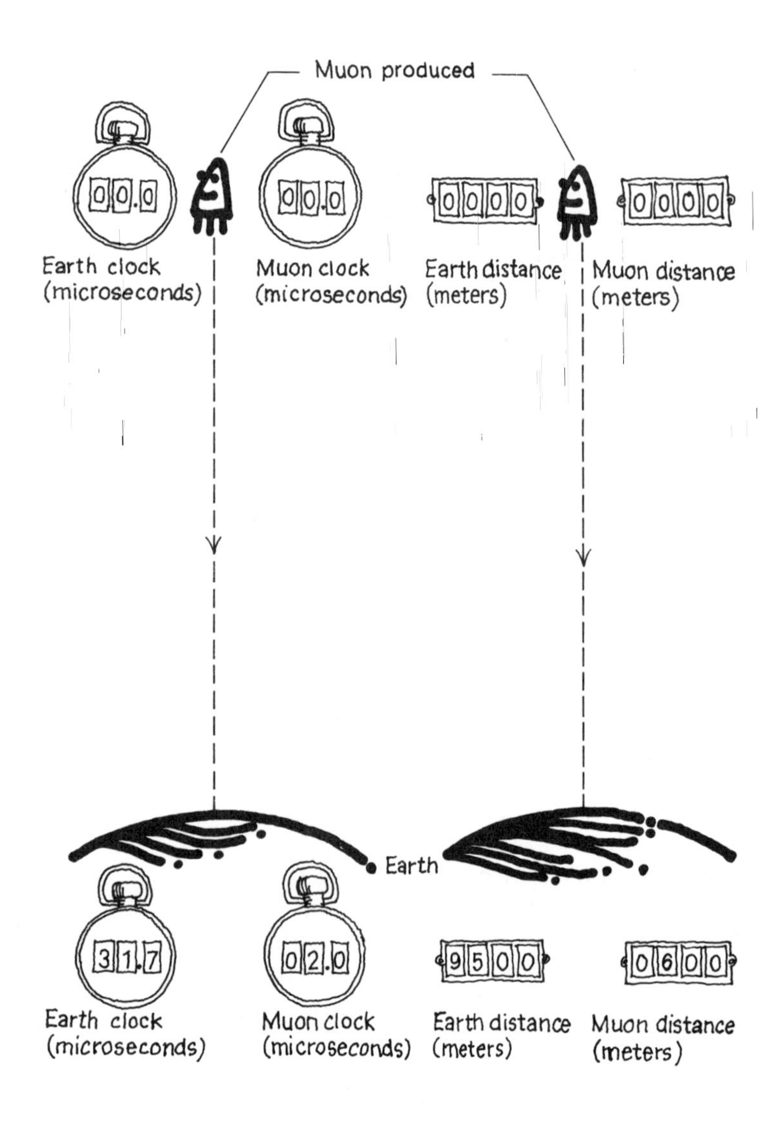

FIG. 3.4 Earth-based observer says muon clock is slow. Muon-based observer says clock is correct; it is only 600 m, not 9500 m, to ground. (A microsecond is 10^{-6} sec.)

dilation effects. When making observations from the frame of reference of the earth, we look at the muon and see that it is moving very rapidly, thus the muon "clock" appears to us to run too slowly (see Fig. 3.4). In fact, when the calculation is made, we find that the muon clock is so "slow" that the 31.7 millionths of a second that we know the muon takes to travel 9500 meters requires only 2 millionths of a second on the muon clock.

Of course the muon doesn't think its clock is slow at all. It believes that it lives 2 millionths of a second, just as it should before decaying, but to the muon the distance it travels is contracted. Again, making the calculation, we find that the muon would measure the 9500 meters of its journey as being contracted to only 600 meters, which is exactly how far a muon can travel in 2 millionths of a second at a speed of 0.998*c*.

3.4 SPACE-TIME

One of the most significant implications of all the things we have been saying is that we can no longer think of space and time as being separate entities. We must realize that we live in a universe in which the three spatial dimensions and the *fourth dimension*, time, are considered together as *space-time*.

For simplicity, let's think for the moment about motion that is in only one space dimension; otherwise we would have to construct a 3- or even 4-dimensional graph. Imagine that you are constrained to move only along a single straight line. We may then represent your position along this one-directional dimension by points on the horizontal axis of a graph. Similarly, moments in time will be

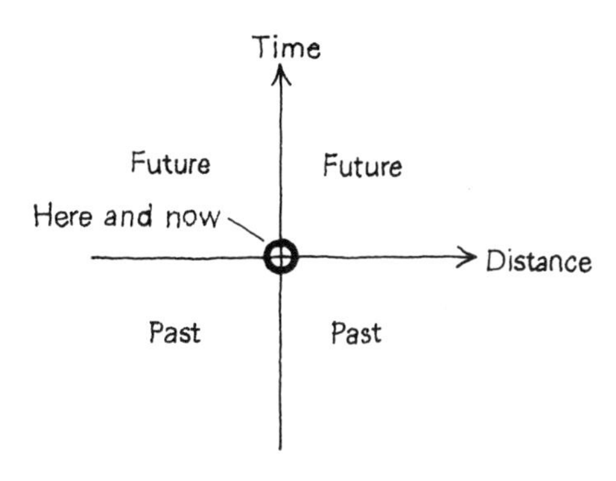

FIG. 3.5 A space-time graph.

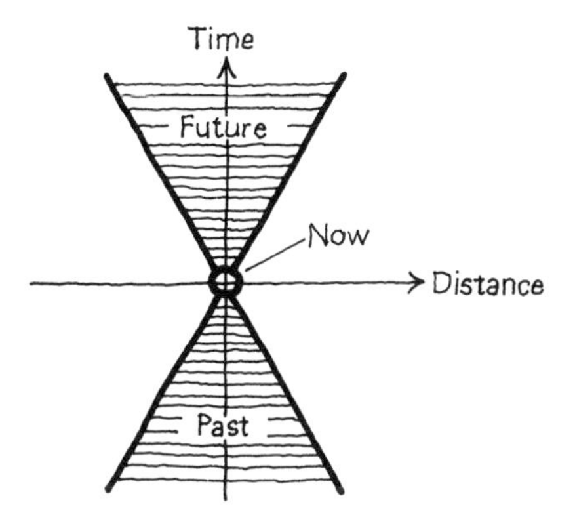

FIG. 3.6

represented by points along the graph's vertical axis. (A space-time graph in one spatial dimension, such as the one we are describing, is shown in Fig. 3.5.) Each point in the plane of this graph represents a unique location in space-time. Just as we may assign a zero point on the space axis and consider positive and negative points on the two sides of this point, the zero on the time axis may represent "now," with points in one direction representing the past and those in the other direction representing the future. To locate events in two spatial dimensions, we would use a plane to represent the space coordinates, and use the third axis for time. To represent points in three spatial dimensions, we would have to construct a 4-dimensional graph.

Because no one can exceed the universal speed limit—the speed of light in free space—once a person's position has been given at any one time, it is possible to specify the limits of where he could have been at any past time or where he can be at any future time. Figure 3.6 illustrates this for travel in one spatial dimension. Point $\oplus$ represents the location of a person "now." The diagonal lines form the boundaries of the places that may be reached by traveling at the speed of light from our starting point. Thus the shaded area includes all the points that may be occupied in the future or that could have been occupied in the past by a person located at "now." This means that there are places in the past where you *could not have been* and places in the future where you *cannot be* simply because you would have to travel faster than the speed of light to reach them. We may therefore conclude that the only events to which you can be *causally connected* are those within the shaded area. This area is known as one's *light-cone.*

Figure 3.6 represents only motion in a single inertial frame for a given observer. A pair of

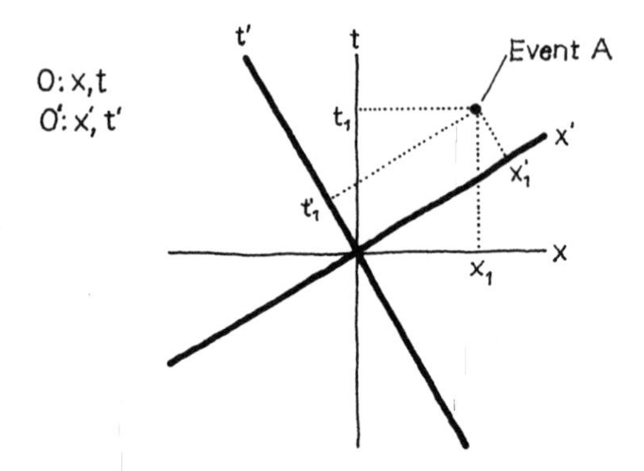

FIG. 3.7

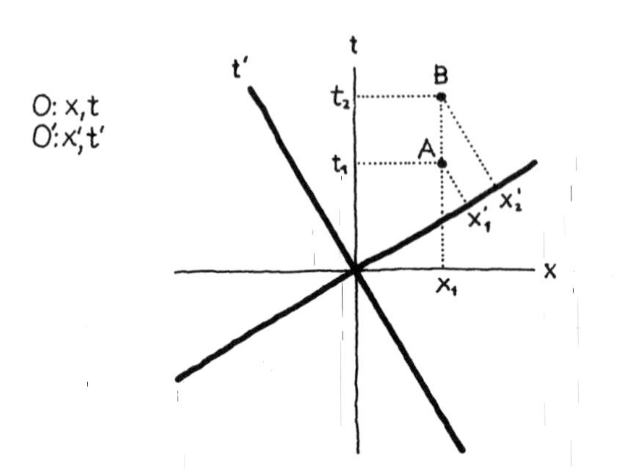

FIG. 3.8 Events A and B occur at the same place but at different times in O, thus are seen at different places in O'.

similar drawings may be used to compare events in two different inertial frames that are moving at constant velocity with respect to each other. The relative motion of the inertial frames is denoted by the fact that one set of axes has been rotated with respect to the other. The angle to which these axes have been rotated is proportional to the relative velocity of either inertial system with respect to the other.

In Fig. 3.7 two observers, let us call them O and O', are located in their inertial systems, moving at a constant speed with respect to one another. We shall concern ourselves now only with motion in the x-direction, so we would say that system O' is moving in the x-direction with constant velocity v with respect to the O system. (And of course system O is moving at constant speed $-v$ with respect to system O'.) Let point A represent some event in space-time. This means that a specific event, which we label A, occurred at a given point in space at a given time. To locate this point as each observer sees it, we simply measure the shortest distance from point A to the space and time axes of the two observers' systems. Observer O says that event A occurred at point x_1 at time t_1, while observer O' says that it occurred at point x'_1 and time t'_1.

There is nothing at all mysterious about the observers measuring different distance co-ordinates for event A. Each observer is simply measuring from his own reference origin to the point at which A took place. Since each observer is located at a different place, each measures a different distance to point A. The two observers also measure different *times* for this single event, but we have already seen that this occurs because the information that informs them of the event can travel to them at only a finite speed.

Now let's make another sort of comparison. Let's compare time and distance *intervals.* To help visualize this situation, let one system simply be the earth, and locate the second

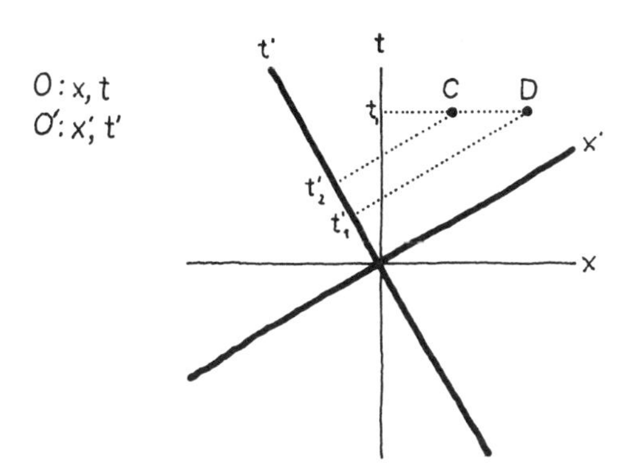

FIG. 3.9 Events *C* and *D* occur at different places but at the same time in *O*, thus are seen at different times in *O′*.

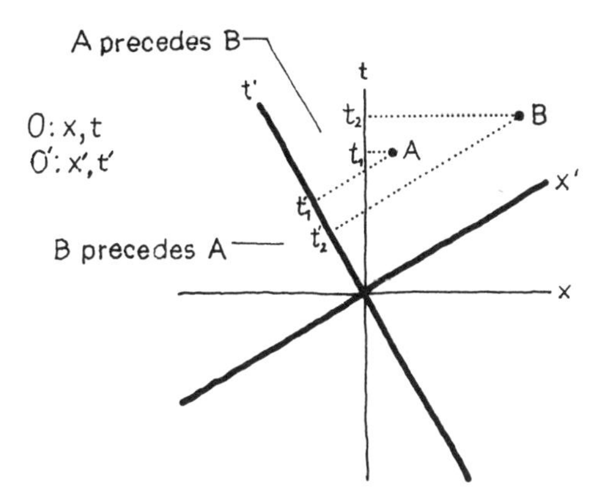

FIG. 3.10 Observers in different inertial systems cannot agree on the *order* in which events took place!

system in a jet airliner passing overhead at a constant speed of 600 miles per hour. Suppose that the stewardess in the airliner serves a passenger his meal and, at some time later, removes the tray. To an observer in the airliner, these events took place at *different times,* but at the *same place* in the coordinate system of the airliner. This follows from the fact that the passenger was seated at the same place when each event occurred. But if our observer on the ground has some means of observing these events, he would disagree, saying that the two events took place not only at *different times* but at *different places.* If the airliner were traveling at 600 miles per hour and the meal required 30 minutes, these events would be separated by 300 miles, as seen by the ground-based observer.

All that this example has illustrated thus far is the perhaps obvious fact that *events that occur at the same place but at different times in one coordinate system will be seen to occur at different places as well as different times in another coordinate system* (see Fig. 3.8). But from our diagram, it may also be shown that the not-so-obvious converse of the statement is also true: *Events that occur at the same time but at different places in one coordinate system will be seen to occur at different times as well as different places in another coordinate system* (see Fig. 3.9). Imagine that two stewardesses in the airliner each enter a cabin through opposite doors at the same time so far as can be told inside the airliner. Because these events occurred at *different places* in the airliner they will be observed from the ground as having occurred at *different times* as well. Thus events that appear simultaneous as seen in one coordinate system will appear to have happened at different times in a coordinate system moving with respect to the first.

Finally, it may be seen from these diagrams that even the *order* in which events took place cannot be agreed upon by observers in different coordinate systems. In Fig. 3.10 it is evident that in system O event A preceded event B, while in system O' the order of the events has been reversed! Again we must point out that, although the effects we have been describing always occur, they are easily observable only if the speed of one coordinate system relative to the other is very near the speed of light.

3.5 THE GENERAL THEORY OF RELATIVITY

In all the preceding discussion, our considerations have been limited to inertial systems that were moving with constant speeds with respect to each other. Now let us consider systems that may be accelerated.

Imagine that you were in a space vehicle somewhere in deep space, with the capsule moving at a uniform speed. You would be experiencing the "weightless" condition that is now familiar to anyone who has seen the TV news films of space flight. Now suppose that the ship were to suddenly undergo a uniform acceleration in a given direction. The sensation you would feel would be the same sort you experience when you are in an automobile that is accelerating rapidly or an airplane that is taking off. If the acceleration were directed toward the "top" of the spacecraft, you would find yourself thrown toward the floor. A better way to describe what would have happened is to say that you stayed where you were and the floor came up to meet you. So long as the upward acceleration continued, you would experience an acceleration toward the floor and could move and walk around on it just as if the space vehicle had developed a gravitational field. You would find that there would be an "up" and a "down" once more.

Now the thing that you must conclude, after a few moments of reflection, is that there is no way of distinguishing between the experience of a constant acceleration in the space vehicle and being in the same vehicle in a gravitational field. Since the acceleration due to gravity on the earth is about 32 feet per second per second, if your spacecraft were to accelerate at a constant 32 feet per second per second, you would weigh the same on a scale placed on the floor of the accelerating spacecraft as you would back on earth.

You might think you could perform some simple physical experiment to distinguish between being accelerated and being in a gravitational field. For example, suppose that you held an object in your hand and released it (Fig. 3.11). In a gravitational field the object would fall downward; in the accelerating space vehicle, the floor would come up to meet it, so that the effect would be the same. There is simply no physical experiment you can perform to distinguish between being accelerated and being in a gravitational field. This is part of what Einstein said in his General Theory of Relativity, first published in 1916.

If there really is no physical experiment that enables you to distinguish between being in a gravitational field and being accelerated, what would happen if you were to shine a beam of light across the cabin of the spacecraft? If the spacecraft is being accelerated, the floor will come up to meet the beam of light, so that the beam appears to curve down toward the floor on the side of the room opposite the light source (Fig. 3.12). Of course, the light beam travels so fast that this effect appears to be quite a small one, but you would nonetheless be in a position to observe a beam of light being "bent" by the acceleration of the space vehicle.

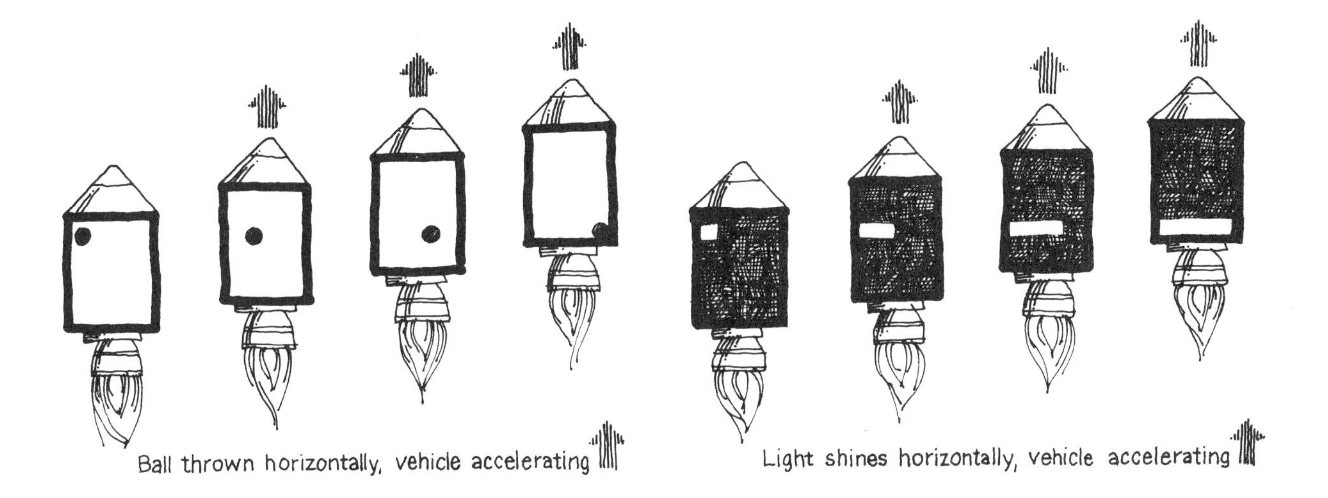

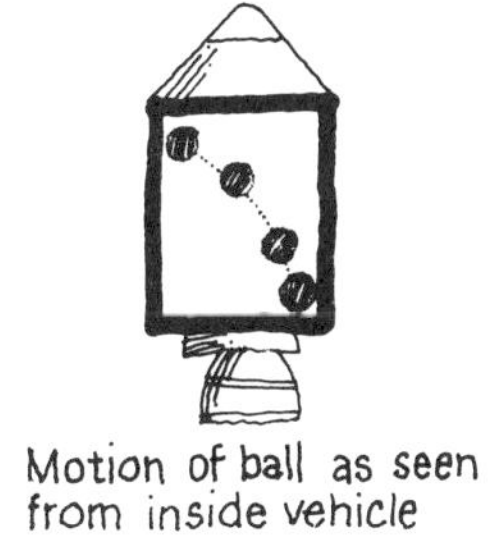

FIG. 3.11

FIG. 3.12

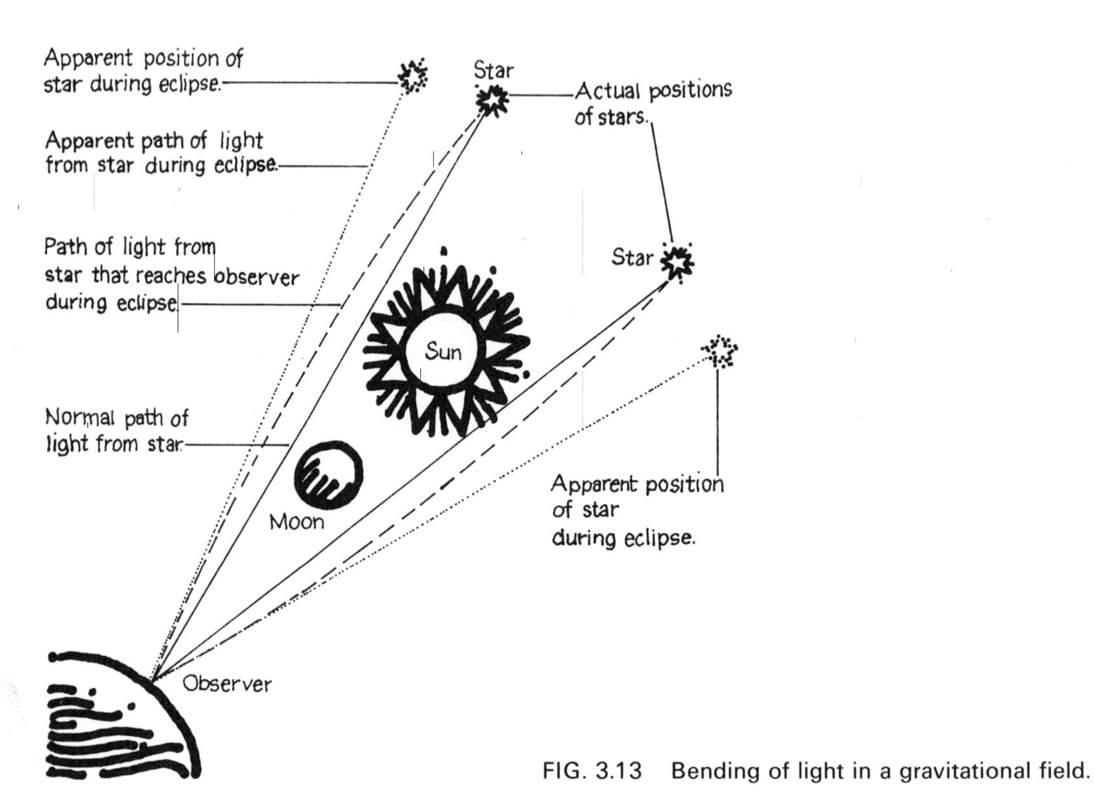

FIG. 3.13 Bending of light in a gravitational field.

But does this also occur when light passes through a gravitational field? If there is no observable difference between being in a gravitational field and being accelerated, a light beam should also be bent by a gravitational field. Thus light would not always travel in a straight line and may even be visibly deflected by a strong gravitational field. In 1919, just three years after Einstein first presented his General Theory of Relativity, there was an opportunity to check on this part of his theory during an eclipse of the sun.

Figure 3.13 illustrates the experiment in which the bending of light in a gravitational field was first observed. Before the eclipse, experimenters measured the positions of two stars. During this measurement, light from the stars had to pass only through free space to reach the earth. During the eclipse, the experimenters again measured the apparent positions of these stars, but now the light from them had to pass very near the sun to reach the observers. The results showed that the stars appeared to be farther from each other during the eclipse than they actually were. The explanation for this effect is that the gravitational field of the sun had *bent* the starlight, thus changing the apparent positions of these stars. We therefore see that light will not travel in a straight line through space, but actually follows a curved path when passing near a massive body such as the sun.

These measurements have been repeated a number of times, but a disadvantage inherent in this experiment is the fact that observations can be made only during total solar eclipses; otherwise the stars will not be visible. However, many astronomical objects emit powerful radio-frequency signals which can be detected after they have passed near the sun, and one can observe them without having to wait for an eclipse. Recent experiments involving radio signals have provided a much more accurate confirmation of this prediction of general relativity than is possible from optical measurements.

A second prediction of the General Theory of Relativity involved orbital motion of planets near massive bodies. In Chapter 2 we found that the perihelion point of the planet Mercury's orbit precesses (the orbit does not close on itself) by 43 seconds of arc per century more than can be explained by Newtonian gravity. This precession is, however, accounted for by general-relativistic calculations.

A third prediction of the General Theory of Relativity is that of the slowing down of a clock near a massive object. This means that a clock runs more slowly at the earth's surface than a few feet above it. This effect, too, has been confirmed experimentally.

Another prediction of the General Theory of Relativity was that gravitational fields are propagated by waves, just as electromagnet interactions involve electromagnetic radiation. For many years the existence of gravitational radiation was debated, but no physical evidence of its existence was found. Then, in June 1969, an American physicist, Joseph Weber, announced that he had detected gravitational radiation.

Weber used a pair of one-and-a-half-ton aluminum cylinders as antennas in his observations. One of the cylinders was located in Maryland and the other near Chicago. Each of these detectors was suspended in a vacuum chamber and isolated from seismic disturbances. Weber interpreted simultaneous responses of these detectors located nearly 700 miles apart as indications that signals due to gravitational radiation had been received and detected.

3.6 THE CURVATURE OF SPACE

As witnessed by all our senses, we clearly live in a world that is 3-dimensional. We conclude this by noting that the position of any object may be uniquely specified by its distances from 3 mutually perpendicular planes. Any corner of a rectangular room is the intersection of 3 mutually perpendicular planes, so any object in the room may be located by its distances from the two walls and floor (or ceiling) intersecting at that corner.

Because all our experiences have been with 3 dimensions, we cannot easily visualize what a space with 4 or more dimensions would be like. To better appreciate this situation, think of the problems a person accustomed to living in a 2-dimensional world might have if he were to try to comprehend the 3-dimensional

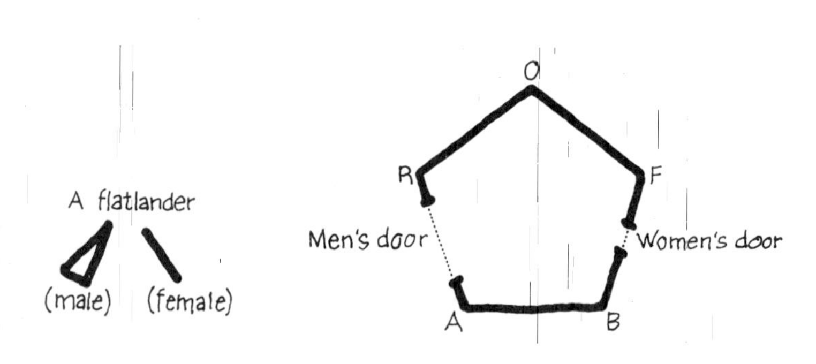

FIG. 3.14 A Flatlander's house (as depicted by Edwin Abbott).

world with which you are familiar. Just such a situation was depicted in a delightful little book, *Flatland,* written by Edwin Abbott. Abbott pictured a world, Flatland, inhabited by 2-dimensional beings (Fig. 3.14). Flatlanders have a finite uniform thickness, though they are unaware of any third spatial dimension. While a Flatlander can perform the usual sort of things any three-dimensional person can, he does them in one less dimension.

A 3-dimensional person could become "visible" in the Flatlander's world by placing a part of himself in the plane of the Flatlander. The Flatlander would see only that part of the 3-dimensional figure that was actually in his plane, thus the apparent shape of the 3-dimensional being could be easily changed—so far as the Flatlander could tell—as his projection in the plane that was Flatland was changed.

The 3-dimensional being would find it simple to do things that could not be understood by the Flatlander, who had no knowledge of a third dimension. For example, though a Flatlander could be closed up in his house with all the doors and windows shut, the 3-dimensional being could place an object inside that house (or take one out) without opening any of its doors or windows. This act that would seem so puzzling to the Flatlander would appear quite commonplace to the 3-dimensional being.

If you still don't appreciate the problems the Flatlander would have, let's look at a few 4-dimensional situations in our 3-dimensional world. Cut out a long strip of paper and bring its ends together as if you were about to fasten them into a loop. But, before joining them, twist one end over (rotate it through 180 degrees). The loop you have made is known as a Möbius strip (Fig. 3.15), and it has the unusual property of being *one-sided.* If you start to trace a line down the center of one side of the strip and continue your mark without stopping until the starting point has again been reached, you will find that you have drawn a line on not one but *both* sides of the surface. If you now try to cut the loop in half by cutting it along this line with a pair of scissors, you will only create two loops that are linked together. You may wish to try to cut a Möbius strip into three sections by cutting around the strip 1/3 of the way in from the edge.

A 4-dimensional being would see nothing strange about any of the preceding situations, for a Möbius strip may easily be untwisted—without its ends being separated—in 4-dimensional space. Similarly, a 4-dimensional

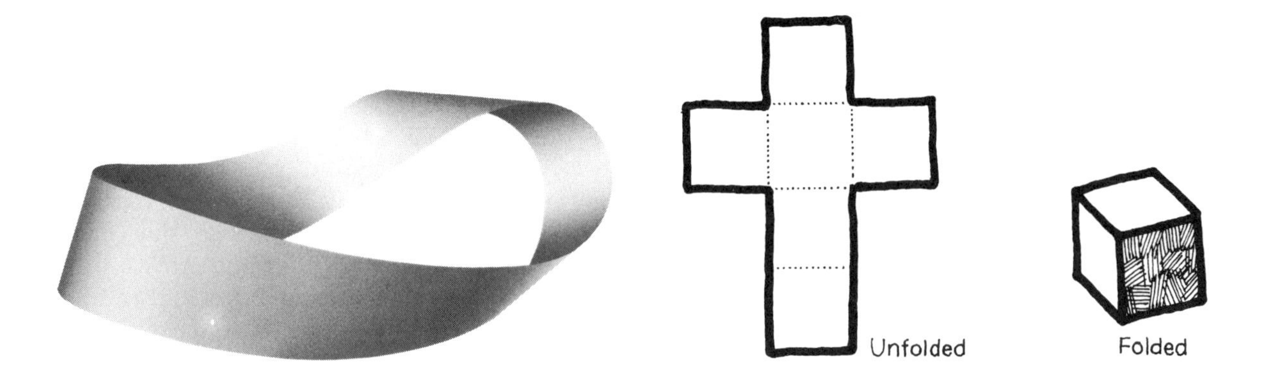

FIG. 3.15 A Möbius strip.

FIG. 3.16 Three-dimensional cube.

being could untie a simple knot that you made in a length of string while you continued to hold on to both ends of the string.

Although we cannot visualize figures in 4 dimensions, we can see what they look like unfolded in 3 dimensions. You are surely familiar with the way in which a piece of paper, cut and folded as shown in Fig. 3.16, may be formed into a cube. You can, of course, see the whole cube in its folded 3-dimensional form as well as in the unfolded 2 dimensional configuration. The Flatlander, though able to see the unfolded cube, could see only a square—the projection—of the folded cube in his 2-dimensional world. Now a 4-dimensional hypercube, unfolded into 3 dimensions, would look like the object illustrated in Fig. 3.17. If it were folded in 4-dimensional space, its 3-dimensional projection would appear as an ordinary cube, but that would only

FIG. 3.17 Four-dimensional cube.

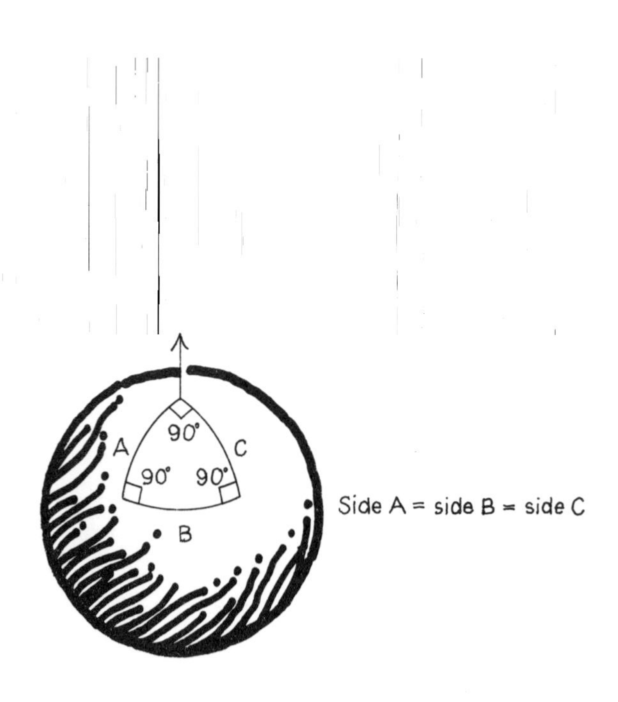

FIG. 3.18

be the projection of part of the 4-dimensional figure in our 3-dimensional space.

Now imagine that you were to put Flatland on a 3-dimensional surface such as a sphere. This would be the equivalent of having Flatlanders living on a globe such as our earth. A Flatlander might experience the mystifying (to him) experience of heading in one direction, walking in a "straight" line, and eventually coming back to the place from which he started. He would, of course, have simply walked all the way around the earth.

The Flatlander learned in his geometry class, as you did, that the sum of the angles of any triangle is 180 degrees. Yet this is not true on a sphere. Starting from the north pole, it is possible to walk a mile due south, make a 90-degree turn and walk a mile due east, make another 90-degree turn and walk a mile due north, to find that you have returned to the spot from which you started (Fig. 3.18). The sides of this triangle are all the same length, so it is an equilateral triangle. Hence the three angles are all the same size: 90 degrees each. The sum of the angles of this triangle is 270 degrees!

Going from one place on the surface of our globe to another via the shortest route involves moving not in a "straight" line, but along a part of a circle. This circle includes the two points between which travel is contemplated, and has its center at the center of the sphere. Such a circle is known as a *great circle,* and the path along it is a *geodesic*. A geodesic is the "straightest" line between any two points; on the surface of a sphere it is a part of a circle.

With these things in mind it is now possible to present a still more general picture of the space-time in which we live. We shall replace the concept of gravitational fields by that of *curvature of the space-time continuum.* All the effects we once attributed to the gravitational field surrounding a massive body may now be explained by saying that the massive body distorts the space-time continuum around it.

To help visualize this situation, imagine a sheet of rubber that has been stretched tightly so that its surface is level. A golf ball rolled

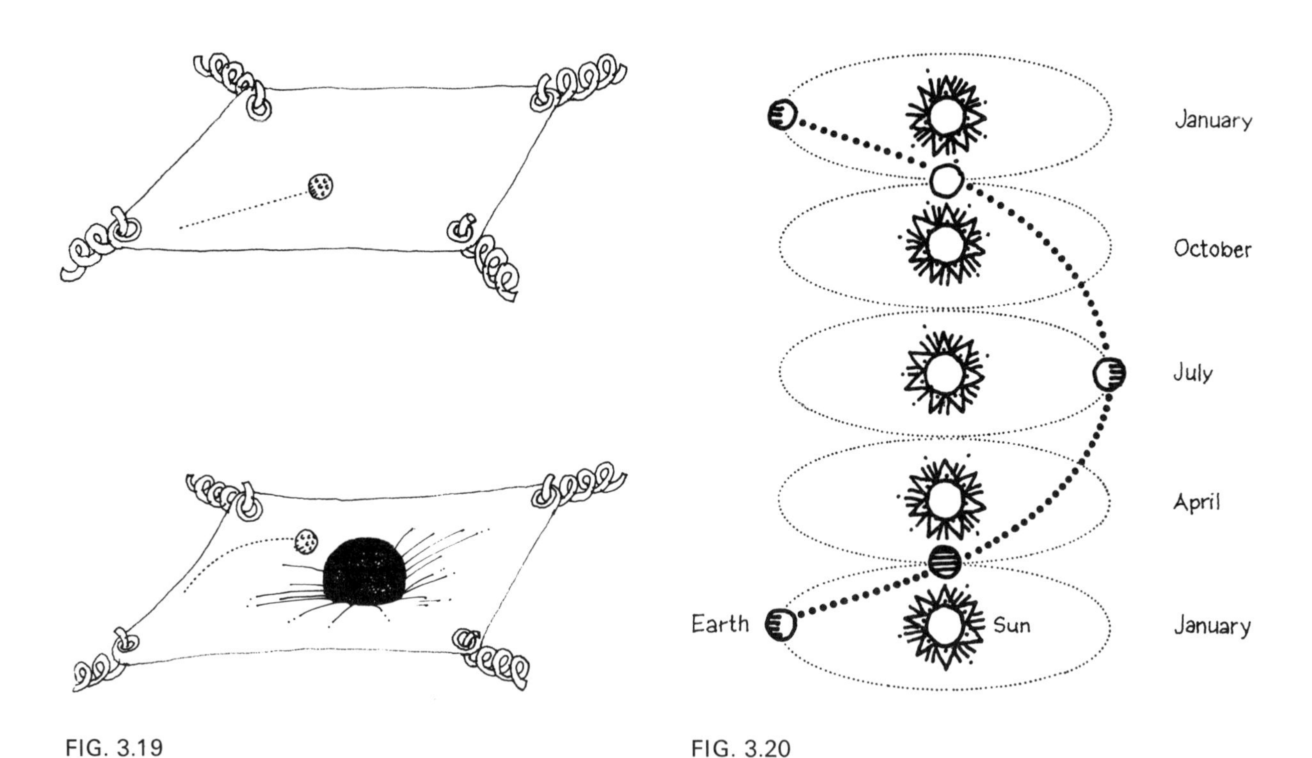

FIG. 3.19

FIG. 3.20

across this sheet will travel in a straight line. But now put a bowling ball in the center of the rubber sheet (Fig. 3.19). The bowling ball will distort the surface of the sheet, so that the path of the golf ball across the sheet will no longer be straight, but curved. The space over which the golf ball rolls has been distorted by the presence of the massive bowling ball. The curvature of the space-time continuum around a massive body is responsible for the attraction of other masses, as well as the bending of rays of light. Thus the facts that planets orbit the sun and that starlight is deflected near the sun are both explained by the curvature of space-time: Planets and starlight both follow geodesics through the space-time near the sun. The path of an object through space-time is called its *world-line*. The earth's world-line during a year is illustrated, in Fig. 3.20, by slices out of space-time.

3.7 THE EXTENT OF THE UNIVERSE

A consideration of the nature of the space in which we live leads, inevitably, to an even more fundamental question: What is the extent of this universe? If the universe is finite in extent, how are its boundaries defined? And what is on the other side of these boundaries? But if the concept of a bounded universe is unsatisfactory, it is even more difficult to think of a universe that is infinite in extent. Or is it possible that our universe is finite but closed upon itself, in a manner similar to a Flatland located on a sphere?

All the information we can receive from the depths of space is derived from the electromagnetic or gravitational radiation we detect at our observatories on or near the earth. Part of this electromagnetic radiation is in the visible region of the spectrum. On any clear night, with the unaided eye, we may see several thousands of stellar objects. Aided by a telescope, the observer is quickly overwhelmed by the number of objects visible. In our galaxy alone there are about 100 billion stars, yet there are some 10 billion other galaxies that may be seen. Even these numbers surely do not indicate the sum total of the objects in the universe, for objects more distant than about 10 billion light years appear too dim to be detected by our largest telescopes.

Significant numbers of other astronomical objects that do not radiate enough energy to be observed in the visible spectrum are detectable at radio wavelengths. It has been suggested that the number of radio galaxies may be comparable in magnitude to the number of visible galaxies. Still other objects emit radiation at frequencies as high as the x-ray region.

Information about the composition of stars may be derived from the study of their spectral lines. Each of the elements in nature, when heated, emits electromagnetic radiation. When this radiation is analyzed with a *spectrometer*—an instrument that separates the radiation by wavelength—one finds that it contains a number of discrete *spectral lines.* A spectral line consists of intense radiation within a very narrow band of wavelengths. The pattern produced by spectral lines of several different wavelengths may be used to uniquely determine the elements that emitted these lines. Performing this type of analysis, one may determine which elements are present in a given star or galaxy.

The analysis of spectra from distant galaxies also reveals that, although the characteristic identifying wavelength patterns of the elements remain the same, their spectral lines are shifted in wavelength with respect to the positions they occupy when the spectra of these same elements are studied in laboratories on earth. This phenomenon, known as *Doppler shifting,* is similar to the effect we observe when we listen to a sound source that is moving relative to us. It is a Doppler shift that is responsible for the fact that the frequency of the wail of the siren on an ambulance appears to rise (wavelength becomes shorter) as the vehicle approaches and lowers (wavelength becomes longer) as the siren recedes. When a source of electromagnetic radiation approaches an observer, the spectral lines it emits shift toward shorter wavelengths (they move toward the blue part of the spectrum). When a source is receding, its spectral lines are shifted to longer wavelengths (toward the red part of the spectrum).

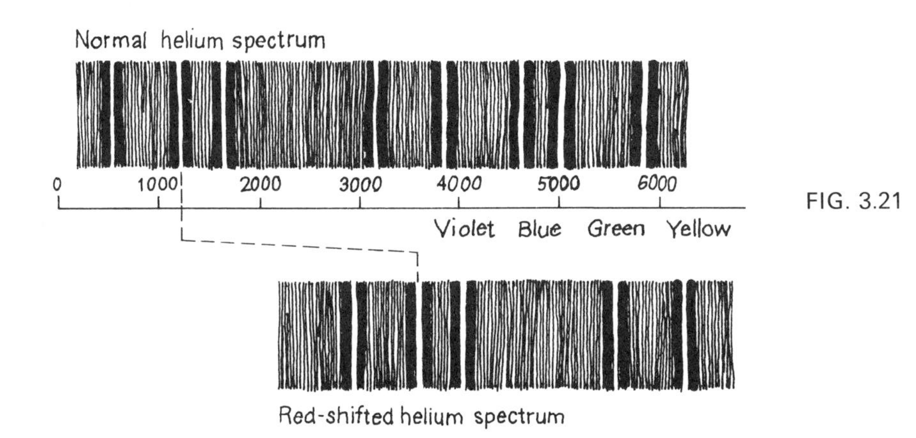

FIG. 3.21

No stellar object has ever been observed to exhibit a shift toward the blue, but radiation from all the galaxies is shifted toward the red part of the spectrum, an effect known as the *red shift* (Fig. 3.21). The fact that red shifts are observed and that blue shifts are not implies that the galaxies are all moving away from one another; the universe is expanding. The most distant galaxies have the greatest red shifts, which indicates that they are moving away from us at the greatest speeds. If one were to calculate backward in time, it would appear that all these objects were once concentrated in the same region of the universe. Following an "explosion," they were scattered, and we are now simply observing that those objects that left the site of the explosion with the greatest speeds have now traveled the greatest distances.

A number of objects discovered in the last few years have almost unbelievably large red shifts. Described at first as quasi-stellar objects, these sources are now familiarly

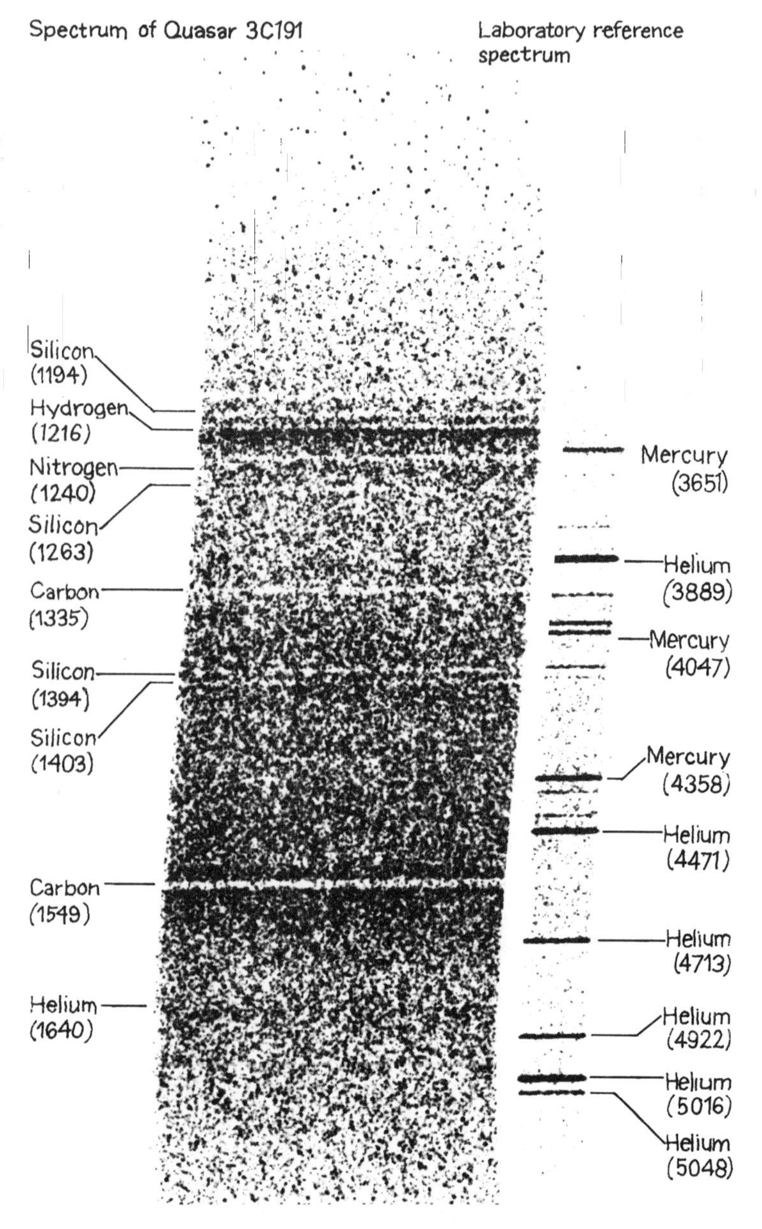

FIG. 3.22 Red-shifted spectrum (normal wavelengths in parentheses)

known as *quasars* (Fig. 3.22). Some quasars show red shifts as great as 200%. If these red shifts are caused entirely by Doppler shifting, then some quasars are traveling at speeds greater than $0.8c$ and must be the most distant objects observable. But from their apparent brightness, it would then follow that quasars must emit energy at a rate many times greater than any non-quasar galaxy, and must utilize an energy-production mechanism different from that of any nearby object in the universe.

FOR MORE INFORMATION

P. A. M. Dirac, "The Physicist's View of Nature," *Scientific American,* May 1963, pages 45–53. An eloquent argument for the importance of symmetry and beauty in physics.

Adolph Baker, *Modern Physics and Antiphysics,* Reading, Mass.: Addison-Wesley, 1970, pages 42–92. A most enjoyable and illuminating discussion of relativistic observations.

Edwin A. Abbott, *Flatland,* New York: Barnes and Noble, 1963. The second revision of this work, published in 1884.

Mendel Sachs, "Space, Time and Elementary Interactions in Relativity," *Physics Today,* February 1969, pages 51–60.

Dionys Burger, *Sphereland,* New York: Crowell, 1965. A fantasy about curved spaces and an expanding universe; includes a condensation of *Flatland.*

George Gamow, *Mr. Tompkins in Wonderland,* Cambridge, England: Cambridge University Press, 1939. A visit to a land where the speed of light is just 20 miles per hour.

Albert Einstein, *The Meaning of Relativity,* Princeton, N.J.: Princeton University Press, 1956.

QUESTIONS

1. How do you suppose a person who had lived his entire life in a closed room on earth would describe the dimensionality of space?
2. In what way did Newton's understanding of gravitation add an extra degree of symmetry to our concept of space?
3. From what two postulates does Einstein's Restricted Theory of Relativity follow?
4. What is the "restriction" in the Restricted Theory?
5. Discuss the statement, "All motion is relative" in terms of two boats on a lake, two airplanes in the earth's atmosphere, two spaceships between the earth and Mars, and two stars in a distant galaxy.
6. A space vehicle traveling away from the earth at half the speed of light passes a similar ship headed for the earth at half the speed of light. Just as the ships pass, each sends a radio message back to earth. Which message arrives first?
7. Describe the problems two relativistic travelers moving at high speed with respect to each other have in making their measurements agree.
8. If the maximum speed with which you can travel is only 1% of the speed of light, how would your light cone compare with the one in Fig. 3.6?
9. Show how two observers in different reference systems could disagree on the *order* in which events they observed took place.
10. Suppose that you were in an elevator when the cables broke and the safety devices failed. As the elevator fell freely toward the earth's surface, is there any experiment you could perform inside the elevator that would give a result different from the same experiment performed in a "weightless" space capsule?
11. Why can the experiment of Fig. 3.13 be performed only during a total solar eclipse?
12. Why could not the gravitational waves detected by Weber have been due to sources within the earth?
13. Sketch a possible arrangement for a Flatland village.
14. What would happen if the men's and women's doors of a Flatland house were both open at the same time?
15. A Möbius strip is sometimes inadvertently made from a belt by a person dressing in a hurry. What could a four-dimensional being do that you can't do to correct this situation?
16. What evidence do we have that the universe is expanding?
17. Why are quasars thought to be so bright?
18. A visitor to a spaceport in the year 2150 witnessed the following scene: A young man emerged from a deep-space rocket and ran to greet a little old lady. As they embraced, she referred to him as her grandfather, yet he appeared to be half her age. Explain.
19. There is some evidence that extended periods of weightlessness produce deleterious effects on the human body. How can this problem be overcome for long space missions?

PROBLEMS

1. If you had been born in an earth-orbiting space station and had lived there until now, how would your view of three-dimensional space differ from that of your friends who were born on earth?

2. As you are accelerated through space to higher and higher speeds, you notice that your wristwatch slows down and then stops. What does this mean?

3. A space traveler, who has been accused of a serious crime, appears in court smiling and confident.

"I have a hundred-percent sure alibi," he tells friends. Can you think what this alibi might be?

4. Discuss the physical implications of the following limerick:

There was a young lady named Bright,
Who could travel much faster than light,
She went off one day
In a relative way
And returned on the previous night.

5. A solution to the following problem should be easy for you. "A hunter, tracking a bear, walked a mile due south, made a 90-degree turn, walked a mile due east, made another 90-degree turn and, after walking one more mile, found he was back at his starting point. What color was the bear?" There is a whole *family* of paths the hunter might have followed that meet the conditions of this problem. Describe them.

6. A physics student was ticketed and brought to trial for running a red light. The student pleaded his case by using relativity to argue that, because he was *approaching* the light, it was shifted toward the blue part of the spectrum and therefore the light appeared *green* to him. The judge, who also knew physics, made a calculation and then fined the student—but not for running a red light. What was the new charge?

7. We have discussed the indistinguishability between being located in an accelerating reference system and being located in a gravitational field. We have also noted that clocks are slowed in gravitational fields. How does the clock of an astronaut who is accelerated compare with a clock on earth? What if the astronaut is *decelerated?*

8. The deep-space ship *Centaurus I* leaves earth for a long mission, during which it will be constantly accelerated for two years. Communications between earth and the spaceship were supposed to have been on the ten-meter band. Yet, as time passes, the earth-based receiver must be tuned to different wavelengths to receive the messages. Why?

9. If observers in three spacecraft positioned about the sun were to point telescopes at each other, would the angles of the triangle they formed appear to add up to 180 degrees?

10. Since light is deflected by a gravitational field, light heading straight for the earth should be attracted by its gravitational field, resulting in "blue-shifted" spectral lines being received on earth. A shift in the positions of lines in the spectrum from the sun has been measured on earth, but this shift is toward the red, not the blue. Why?

11. The existence of tachyons (particles able to travel faster—but never slower—than light) has been speculated. What changes would be necessary in our concepts of making relativistic measurements if tachyons should exist and be detectable by *tardons* (anything slower than light)?

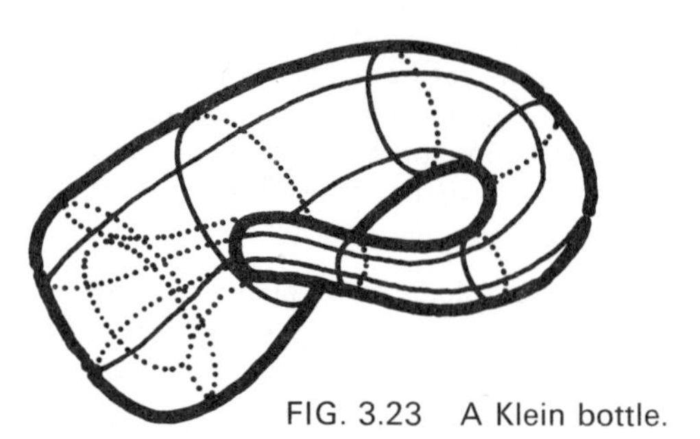

FIG. 3.23 A Klein bottle.

12. A closed Möbius strip is called a *Klein bottle* (Fig. 3.23). Discuss the usefulness of a Klein bottle as a container for liquids, and think of any unique advantages or disadvantages it might have.

13. If there were a fourth spatial dimension in our universe, what "inexplicable" phenomena could we observe that would be associated with this extra spatial dimension?

14. It was once charged that the study of relativity is as useless as the study of the most abstract mathematics or esoteric philosophy. Discuss this point of view.

CHAPTER 4 THE AGE OF THE BOMB

4.1 A DECISIVE MOMENT

At about 5:30 in the morning of July 16, 1945, near a place in the New Mexico desert called Alamogordo, the world's first man-made nuclear device was detonated. Approximately two pounds of plutonium exploded with the destructive force of 20,000 pounds of TNT. The world had entered the nuclear age.

Two nuclear bombs, one using plutonium and one uranium, were dropped on Japan during World War II. Since that time the number of nations that have built bombs has been increased to at least five, the vastly more powerful H-bomb has been developed, and our civilization has lived under the threat of a nuclear holocaust that could destroy civilization as we know it.

Yet nuclear devices have also been used for peaceful applications, and the almost unlimited power available from nuclear sources can also transform our world into something very different and vastly better than the one we now know.

4.2 THE IDEA OF THE ATOM

To develop the vocabulary necessary to intelligently discuss either bombs or peaceful applications of nuclear power, we must reach back as far as ancient Greece. Imagine taking a sample of some material, such as a bar of iron or copper, and dividing it in two; then dividing each half into two more pieces, and these halves yet again, and continuing the process on and on. You would finally be faced with the question of whether or not there is a limit to this sort of division: Is there some smallest size of matter, beyond which we cannot go? That question is at least as old as the ancient Greek civilization, and two of their thinkers—Leucippus and Democritus—answered it by saying that all matter is composed of discrete particles which cannot themselves be subdivided. And they gave us the word "atom," from the Greek *atomos*, meaning indivisible.

But other than knowing that atoms must be quite small and supposedly indivisible, mankind knew almost nothing else about them until the start of our own century. Electrical experiments had disclosed that there are two types of electric charge, which we simply call positive and negative. In 1897 Sir J. J. Thomson had discovered the particle that carries the fundamental unit of negative charge, the *electron*. One expects the normal atom to be electrically neutral—that is, to have equal amounts of positive and negative charge—so a model of the atom, now known as the

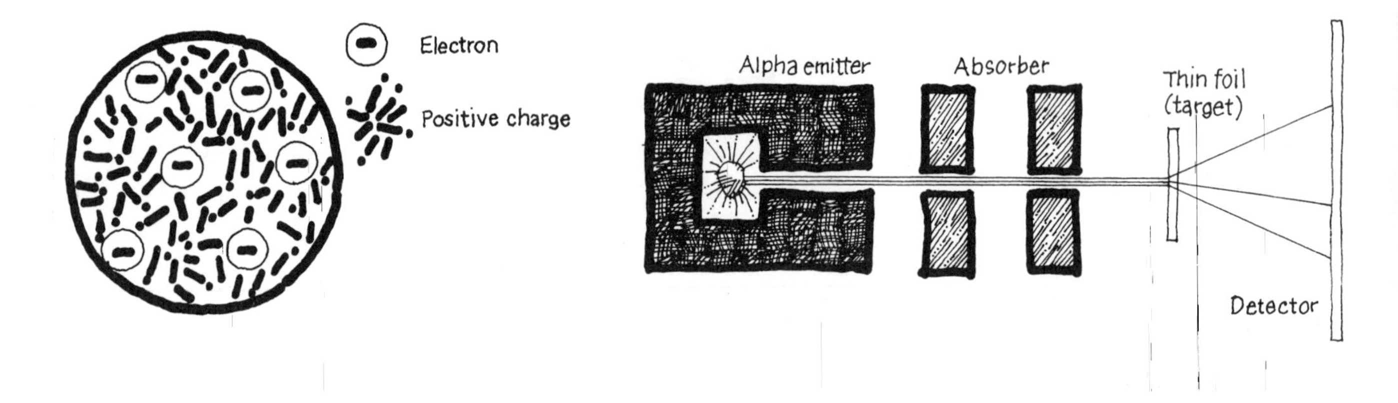

FIG. 4.1 Plum-pudding model of the atom.

FIG. 4.2 Alpha-particle-scattering experiment.

"plum-pudding model," was suggested by Thomson (Fig. 4.1). Thomson pictured the whole of the atom as being positively charged, with the electrons distributed throughout it like plums in a pudding.

This picture had to be radically modified after a series of experiments initiated in 1911 by a former New Zealander working in England, Ernest Rutherford. Rutherford sought to find out something about the size of the atom by an experiment in which small particles were scattered by the atom. By observing the directions in which the particles were scattered, he expected to be able to make estimates of the size and shape of the atom, the object responsible for the scattering.

Rutherford had to use particles of atomic size or smaller to make the measurements meaningful. Fortunately, particles suitable for such experiments are emitted spontaneously by naturally radioactive materials. One particle that is emitted by certain of these materials is called the *alpha particle*. It was known that alpha particles carry positive electrical charges; thus if these particles were allowed to bombard an atom, one would expect them to be scattered by the positive electrical charges of the atom.

To carry out his experiment, Rutherford and his coworkers, Hans Geiger and Ernest Marsden, used a setup like the one shown in Fig. 4.2. The radioactive material emits alpha particles in all directions. Therefore Rutherford placed a heavy shield between the alpha emitter and the target to stop all particles except those that went through a small hole in this shield. This arrangement allowed for the definition of a narrow *beam* of particles. For targets, Rutherford used very thin foils of materials such as gold or copper or silver. The target foil was made very thin so that an alpha particle passing through it would encounter as few atoms as possible. The sort of particle

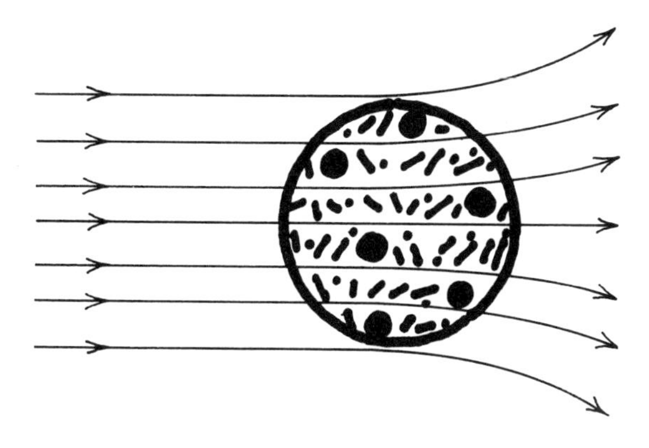

FIG. 4.3 Scattering from plum-pudding atom.

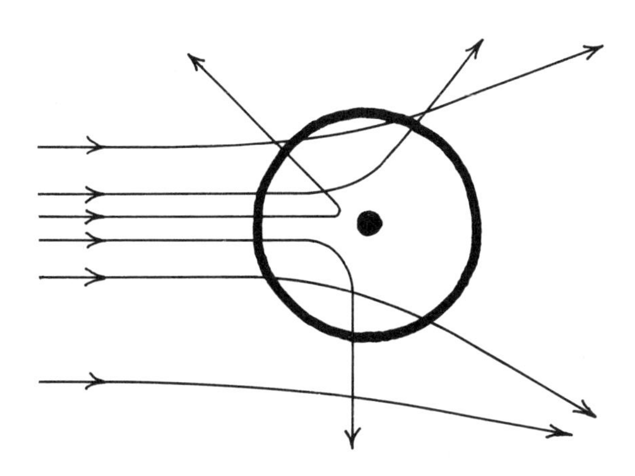

FIG. 4.4 Scattering actually observed.

scattering that Rutherford expected to see in these experiments would involve the alpha particles being deflected from their initial paths through only small angles (Fig. 4.3). Instead, he saw that, although some of the particles were deflected at small angles, others were deflected at very large angles, and some of them were scattered back in the direction from which they had come (Fig. 4.4)! Rutherford and his coworkers were astounded at the large angles through which the alpha particles were scattered. He later wrote of his reaction to this discovery:

It is the most incredible event that has ever happened to me in my life. It was almost as incredible as if you fired a 15-inch shell at a piece of tissue paper and it came back and hit you.

The only way in which this surprising result could be interpreted was to introduce a new concept: the *atomic nucleus,* a tiny area at the center of the atom that contained all the positive charge. It was the repulsive effect of this great concentration of positive charge on the positively charged alpha particles that was responsible for the large-angle scattering observed. In his measurements Rutherford was able to estimate the size of the nucleus as being about 5×10^{-13} cm across. Thus the nucleus is so small that if you could put them side by side, you could put 25,400 billion of them in a space an inch long (Fig. 4.5).

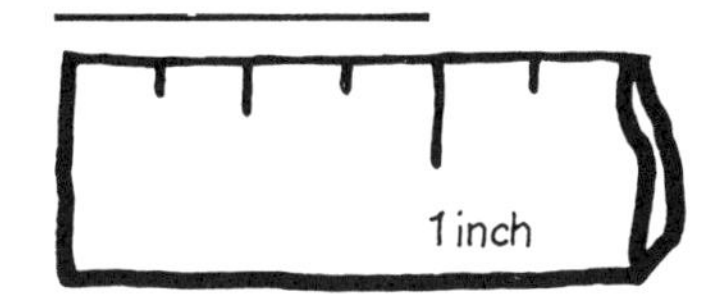

FIG. 4.5

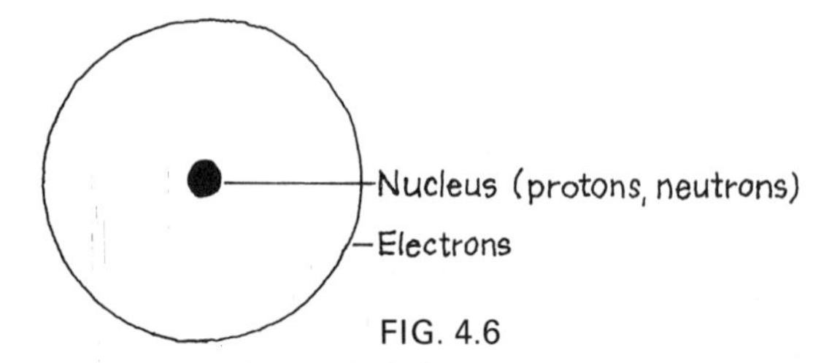

FIG. 4.6

Just as negative electric charge is carried by the electron, positive charge resides in the *proton*. The proton carries an amount of positive charge exactly equal to the amount of negative charge carried by an electron, though the proton has about 1836 times the mass of an electron.

The picture of the atom underwent another important modification in 1932, following the discovery of the *neutron* by Sir James Chadwick. The neutron, a particle that is the same size as a proton, also resides in the nucleus of the atom, but has *no* electric charge.

Thus we have arrived at a simple picture of the atom, in which there is a central nucleus containing protons and neutrons, with the atom's electrons orbiting this nucleus (Fig. 4.6). In any neutral atom, the number of protons is equal to the number of electrons, and this number is the chemical signature of the atom. For example, there are 79 protons in the nucleus of the atom we know as gold. It is because there are precisely 79 protons in this nucleus and 79 electrons orbiting about it that this atom possesses those characteristics we are familiar with as being those of gold. Each element has a different number of protons (or electrons) and this fact allows us to make an orderly arrangement of the elements we find in nature in a *periodic table* (Fig. 4.7).

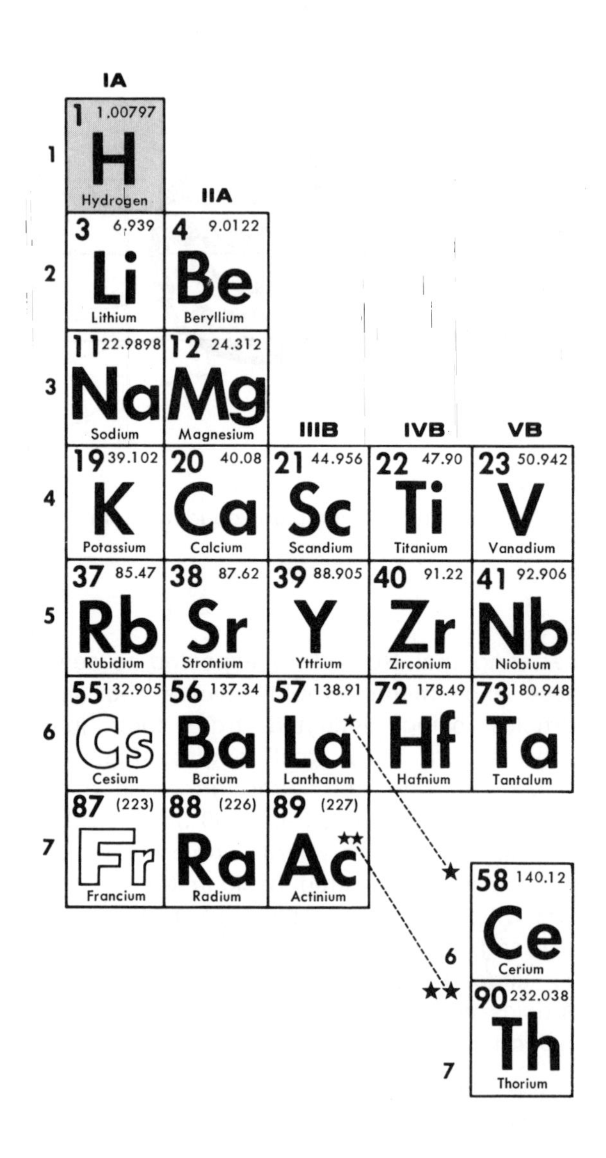

FIG. 4.7 The periodic table of the elements.

VIB	VIIB	VIII	VIII	VIII	IB	IIB	IIIA	IVA	VA	VIA	VIIA	O
												2 4.0026 He Helium
							5 10.811 B Boron	6 12.01115 C Carbon	7 14.0067 N Nitrogen	8 15.9994 O Oxygen	9 18.9984 F Fluorine	10 20.183 Ne Neon
							13 26.9815 Al Aluminum	14 28.086 Si Silicon	15 30.9738 P Phosphorus	16 32.064 S Sulfur	17 35.453 Cl Chlorine	18 39.948 Ar Argon
24 51.996 Cr Chromium	25 54.938 Mn Manganese	26 55.847 Fe Iron	27 58.933 Co Cobalt	28 58.71 Ni Nickel	29 63.54 Cu Copper	30 65.37 Zn Zinc	31 69.72 Ga Gallium	32 72.59 Ge Germanium	33 74.922 As Arsenic	34 78.96 Se Selenium	35 79.909 Br Bromine	36 83.80 Kr Krypton
42 95.94 Mo Molybdenum	43 (98) Tc Technetium	44 101.07 Ru Ruthenium	45 102.905 Rh Rhodium	46 106.4 Pd Palladium	47 107.870 Ag Silver	48 112.40 Cd Cadmium	49 114.82 In Indium	50 118.69 Sn Tin	51 121.75 Sb Antimony	52 127.60 Te Tellurium	53 126.904 I Iodine	54 131.30 Xe Xenon
74 183.85 W Wolfram	75 186.2 Re Rhenium	76 190.2 Os Osmium	77 192.2 Ir Iridium	78 195.09 Pt Platinum	79 196.967 Au Gold	80 200.59 Hg Mercury	81 204.37 Tl Thallium	82 207.19 Pb Lead	83 208.980 Bi Bismuth	84 (210) Po Polonium	85 (210) At Astatine	86 (222) Rn Radon

59 140.907 Pr Praseodymium	60 144.24 Nd Neodymium	61 (147) Pm Promethium	62 150.35 Sm Samarium	63 151.96 Eu Europium	64 157.25 Gd Gadolinium	65 158.924 Tb Terbium	66 162.50 Dy Dysprosium	67 164.930 Ho Holmium	68 167.26 Er Erbium	69 168.934 Tm Thulium	70 173.04 Yb Ytterbium	71 174.97 Lu Lutetium
91 (231) Pa Protactinium	92 238.03 U Uranium	93 (237) Np Neptunium	94 (242) Pu Plutonium	95 (243) Am Americium	96 (245) Cm Curium	97 (249) Bk Berkelium	98 (251) Cf Californium	99 (254) Es Einsteinium	100 (255) Fm Fermium	101 (256) Md Mendelevium	102 (254) No Nobelium	103 (257) Lw Lawrencium

ATOMIC NUMBER	ATOMIC WEIGHT
SYMBOL	
NAME	

Explanation:

Symbol

BLACK — solid

DARK GRAY — synthetically prepared

LIGHT GRAY BLOCK — gas

OUTLINED — liquid

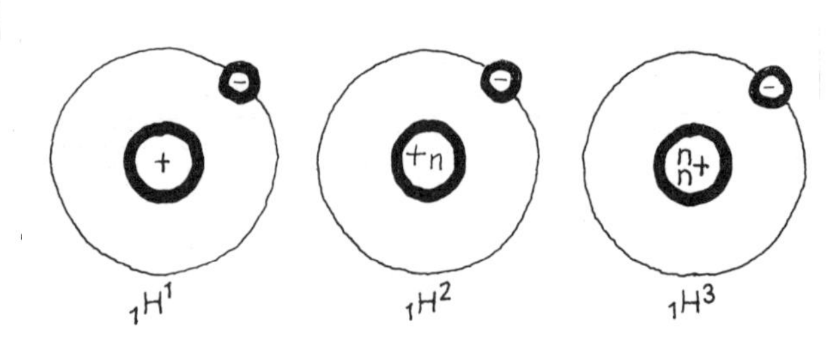

FIG. 4.8

Ninety-two elements are found in nature, each containing from 1 to 92 protons in the nucleus. Hydrogen, the lightest element, has 1 proton, and uranium, the heaviest naturally occurring element, has 92 protons. Although the chemical properties of an element depend on the number of electrons (and therefore protons) it has, different nuclei of any given element may contain different numbers of neutrons and therefore have different physical properties. These *forms of a single element* that have the same numbers of protons but different numbers of neutrons are known as *isotopes* of the element.

We often use a type of shorthand notation to represent elements. This notation is illustrated in Fig. 4.8. The isotopes shown are those of hydrogen (H). The subscript to the left of the letter is the number of protons in the nucleus. Since the element is hydrogen in each case, this number is 1 for all three isotopes. The letter itself is the symbol for the element; therefore using both the subscript and the letter is redundant, so both are not always written. The superscript to the right of the letter represents the number of protons plus the number of neutrons in the nucleus. The three isotopes of hydrogen shown have zero, 1, and 2 neutrons, respectively. The isotope H^2 is known as *deuterium* and H^3 as *tritium*.

The alpha particle that Rutherford used in his scattering experiments was actually the nucleus of the helium atom, containing two protons and two neutrons, He^4.

4.3 NUCLEAR REACTIONS

As early as 1919, Lord Rutherford had bombarded nitrogen nuclei with alpha particles and had obtained oxygen nuclei as a product of the reaction. What was involved was a *nuclear transmutation:* one element had been changed into another by a nuclear reaction. But reactions of this type are difficult to bring off using only alpha particles emitted by natural emitters, since the positively charged nuclei will repel the positively charged alpha particles. In the case of the heavier elements (more protons in the nucleus) these reactions cannot take place unless the approaching alpha particle has enough energy to overcome the repulsive electrical force. Today high-energy particle accelerators make it possible for alpha—and other—particles to reach the energies necessary to do this, but these machines were not available until several years after the time of Rutherford's work.

However, the neutron, carrying *no* electric charge, has no problem in easily approaching

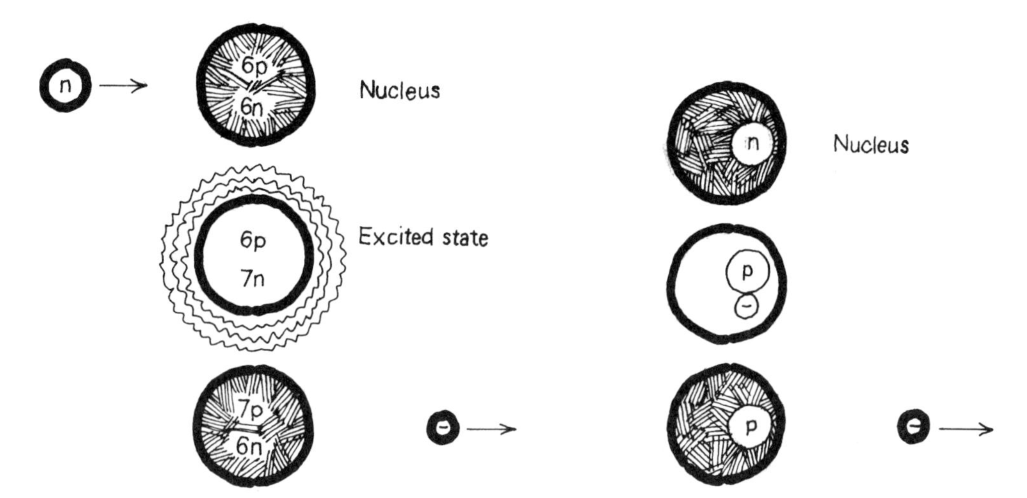

FIG. 4.9 Neutron capture, electron emission.

FIG. 4.10 Beta decay.

and being captured by a nucleus. A nucleus that has just captured a neutron is said to be in an *excited state*, since it possesses more particles—and more energy—than it had found comfortable before (Fig. 4.9). To get out of this excited state, the nucleus *decays* by emitting a particle. It can emit a neutron (not necessarily the same one that it captured) and remain the same element that it was before. Another way in which this decay may take place is by the emission of an electron (Fig. 4.10), in a process known as *beta decay*.

What happens inside the nucleus during beta decay is that a neutron is changed into a proton plus an electron. (Do not get the mistaken idea that a neutron is composed simply of a proton plus an electron for, given sufficient energy, a proton can decay into a neutron plus a positive electron, or *positron*.) The nucleus that remains after undergoing beta decay then contains one more proton than it did before, and it is therefore a different element: one that has a proton number that is 1 greater than the number of protons of the element with which the reaction began.

Enrico Fermi, an Italian physicist, was quick to recognize the advantages of using neutrons to initiate nuclear reactions. In 1934 he began a systematic bombardment of each of the elements with neutrons, starting with the lightest—hydrogen—and working his way up through the periodic table.

In the course of this work Fermi made a most significant discovery: A neutron is much more likely to be captured by a nucleus if it is moving slowly than if it is traveling at a great speed. But how does one slow neutrons down? Since they have no electric charge, neutrons obviously must lose energy through collision with other subatomic particles.

Not every type of collision a neutron makes will slow it down very much, however. Think of what happens when a billiard ball collides with a massive object (the table, for example). The ball rebounds at almost the same speed it had before the collision. Yet in a head-on collision with an object its own size—another billiard ball—it stops, having lost all its energy in a single collision. Hence it is obvious that if one wants slow neutrons, one gets them to collide with particles as nearly as possible their own size. Hydrogen nuclei—protons—have just about the same mass as neutrons. Thus one of Fermi's first neutron moderators was paraffin, a material that contains lots of hydrogen. This worked so well that Fermi decided next to try water as a moderator. He proceeded to perform this experiment in the closest sizable body of water he could find, which happened to be a goldfish fountain behind the physics building at the University of Rome.

Finally, Fermi and his group bombarded the heaviest of the natural elements, uranium, with neutrons. If uranium should capture a neutron and decay by beta emission, the element that remained would be number 93, a *synthetic element* not found in nature. If this process were to be repeated, element number 94 would be formed.

This is indeed what happens. These first two synthetic elements are named *neptunium* and *plutonium*, respectively, after Neptune and Pluto, which are the next planets beyond Uranus in our solar system. Synthetic elements have now been created in laboratories up through atomic number 105. Names given them thus far are listed in Table 4.1.

TABLE 4.1

Atomic number	Symbol	Element
93	Np	Neptunium
94	Pu	Plutonium
95	Am	Americium
96	Cm	Curium
97	Bk	Berkelium
98	Cf	Californium
99	Es	Einsteinium
100	Fm	Fermium
101	Md	Mendelevium
102	No	Nobelium
103	Lw	Lawrencium
104	—	—
105	(Ha)	(Hahnium)

The bombardment of uranium by neutrons did produce some very exciting results for, after the reaction, Fermi and his group found that their sample contained another element which clearly was not uranium. Fermi may have produced element number 93, but he almost certainly had performed an experiment of far greater importance.

The full story of what happened to that uranium could not be told until 1938. In that year Otto Hahn and Fritz Strassmann, in Germany, were bombarding uranium samples with neutrons, seeking to measure the chemical properties of neptunium and plutonium, when —quite unexpectedly—they found a trace of the element barium in their samples. These samples should have contained only uranium

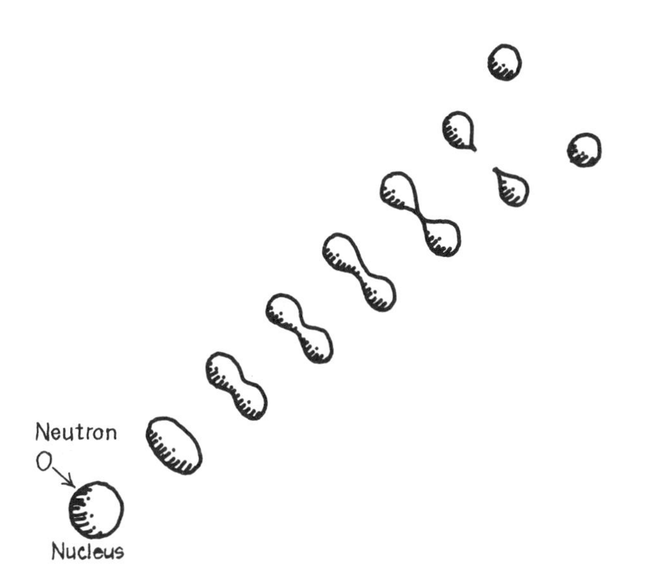

FIG. 4.11 Liquid-drop model of nuclear fission.

FIG. 4.12

and heavier elements. What was a light element such as barium (56 protons) doing among them? The answer to that question has changed the course of history.

The barium was in Hahn's samples because the uranium nucleus had split in two: it had undergone *fission* (Fig. 4.11). The significant thing about this fission process is that the uranium nucleus plus the neutron it captures is heavier than the barium plus the other light nucleus—for example, krypton—plus the neutrons that are given off in the reaction (Fig. 4.12). The nuclear masses involved in this reaction are recorded on the ledger sheets shown in Fig. 4.13. Note that the two sides of the ledger do not balance! Nature's books are never left unbalanced, for the mass that

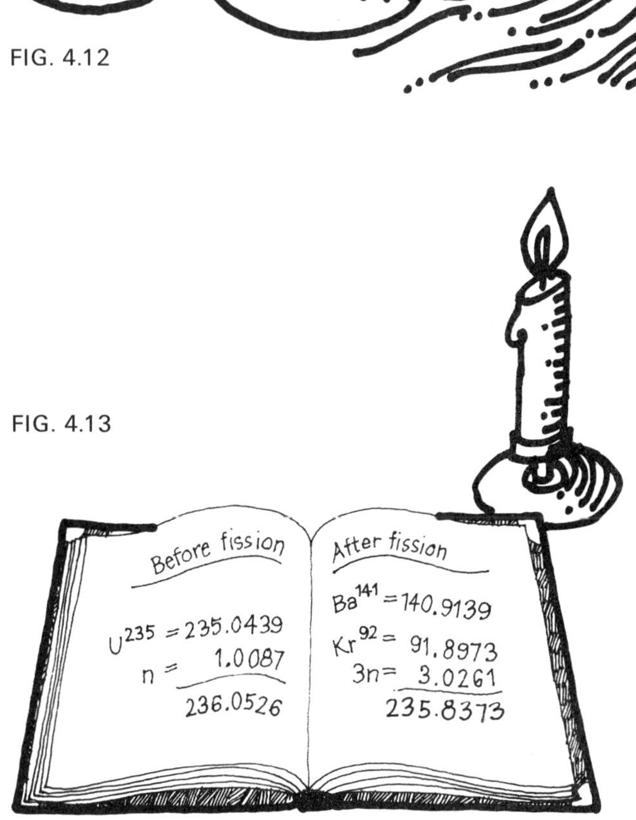

FIG. 4.13

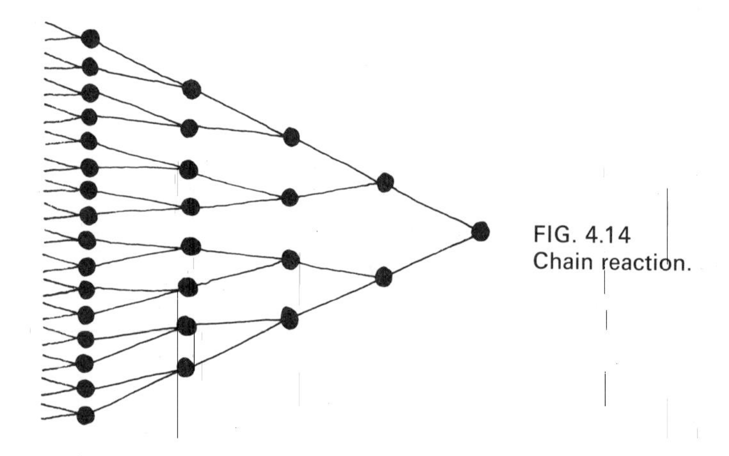

FIG. 4.14
Chain reaction.

is missing on the right has been transformed into energy. You have seen the famous Einstein equation, $E = mc^2$. Now you are seeing what this equation is all about: the missing mass has been released as *energy*.

The amount of energy that is released with the fission of a single uranium nucleus is very small: about enough energy to heat a thousandth of an ounce of water a millionth of a degree. But nuclei are very small, and there are 7.2×10^{22} uranium nuclei in a single ounce of uranium. If enough uranium nuclei could be made to undergo fission in a short time period, the energy released would be enormous.

In our illustration we have shown three neutrons emitted in the fission process. On the average about 2.5 neutrons are emitted for each uranium nucleus that is split. Thus it is possible to imagine an arrangement in which the splitting of a single uranium nucleus would produce fission in two other nuclei, which in turn would lead to the fission of 4, then 8, then 16, in a *chain reaction* (Fig. 4.14). If such a process were to take place rapidly and in an uncontrolled fashion, one would have the makings of a bomb.

4.4 BRAINS TO MAKE THE BOMB

In the 1920's and 1930's, almost all the important physical scientists in the world lived and worked in Europe. After receiving his Ph.D. from an American university, a physicist was not considered fully trained until he had spent a period of time working at one of the great laboratories in Europe. All the important discoveries we have described in the preceding sections were made in England, Germany, Italy, and France.

But the political climate of Europe changed rapidly during the 1930's. In particular those scientists of Jewish ancestry were made to feel unwelcome in Germany and Italy. Einstein had his work attacked as "Jewish physics," and he came to live and work in the United States in 1933. Others in Germany had sought to defend their right to live and work in their native land. The mathematician Courant, who had been wounded and decorated as a German soldier in World War I, found that even this did not prevent him from being stripped of his position and imprisoned.

Similar measures were soon taken in Italy. Enrico Fermi was not a Jew, but his wife was. In November 1938, a new series of strict racial restrictions was announced by the Italian government: Jews were to be deprived of full citizenship rights and their passports would be withdrawn, leaving them virtually prisoners in their own country. A phone call from Stockholm changed all that for the Fermis: Enrico had won the Nobel prize for his work—in the goldfish fountain among other places—with slow neutrons. Fermi was allowed to take his family with him from Italy to Stockholm and thence to the United States.

The list of physical scientists who were driven from the Continent or who left on their own in protest is too long to record here, but many of their names were to become important in the story of the making of American nuclear devices.

4.5 THE NEWS IS OUT

Among those persons driven from Germany by the Nazi regime were two of Hahn's co-workers, Lise Meitner and her nephew, Otto Frisch, who had gone to Denmark to work with the renowned Niels Bohr. They told Bohr of Hahn's and Strassmann's experiments, and correctly deduced that they had observed nuclear fission. An experiment was then designed to test the fission hypothesis, and Bohr left for a visit in the U.S.

On January 26, 1939, Bohr attended a conference of physicists at George Washington University. There he revealed that Lise Meitner had telegraphed him the news that her experiments confirmed the fission theory. His announcement of this discovery electrified the conference and many of those in attendance did not remain for the rest of the proceedings, but rushed back to their laboratories to start fission experiments.

By the middle of February 1939 at least 4 laboratories in the U.S.—those at Columbia, Johns Hopkins, Berkeley, and the Carnegie Institute—had verified Hahn's work. The September 1, 1939, issue of *Physical Review* contained an article by Bohr and John A. Wheeler of Princeton, called "The Mechanism of Nuclear Fission." And by December of that year nearly a hundred papers on fission had appeared.

On learning of the experimental verification of fission, some physical scientists' first reaction was to attempt to enlist all the world's nuclear scientists in a conspiracy of silence. These men knew that it would probably be possible to construct a bomb of hitherto unimagined destructive power, but they also knew that at that time there were perhaps fewer than two dozen physicists in the world who had sufficient understanding of the process involved to actually make such a bomb. The suggestion made was that all the physical scientists of all the world should simply refuse to cooperate with their various governments in the production of nuclear weapons.

Other scientists argued that their deliberate withholding of information or refusal to cooperate in experimentation was unwise, since there was no possible way of knowing in advance what the outcome of their work might be. The results might be very different from what they had imagined. Might it not be possible that a very beneficial device could instead be constructed?

Ultimately, virtually all scientists agreed that, in view of the actions of the Nazi regime, the most reasonable course for scientists of the free world was to proceed with whatever experimentation and development was necessary and possible—if for no other reason than to prevent Hitler from producing the bomb first.

4.6 THE MANHATTAN PROJECT

After confirmation of the discovery of uranium fission reactions in Germany in 1938, it was obvious to physicists on both sides of the Atlantic that a fission bomb might be possible.

Some scientists expressed doubts that a self-sustaining uranium chain reaction could ever be achieved. However, if this *were* possible, the working principles of a bomb seemed clear. Only the details of the actual construction remained to be worked out.

Because a lid of secrecy had been clamped on news coming from Germany, the fears of those persons in America who knew of the potential of such a bomb rapidly mounted. In the summer of 1939, two exiled Hungarian physicists, Leo Szilard and Eugen Wigner, sought out Albert Einstein and persuaded him to write a letter to President Franklin Delano Roosevelt urging the immediate entry of the United States into the race to construct a nuclear device. The letter was actually conveyed to the president by a trusted advisor, New York financier Alexander Sachs, who counted among his personal achievements a more-than-amateurish knowledge of physics.

Einstein's letter to President Roosevelt conveyed the ominous implications of what the German government might do. It began, "Some recent work by E. Fermi and L. Szilard . . . leads me to expect that the element uranium may be turned into a new and important source of energy in the immediate future. . . . This new phenomenon would also lead to the construction of bombs, and it is conceivable . . . that extremely powerful bombs of a new type may thus be constructed." The letter concluded:

I understand that Germany has actually stopped the sale of uranium from the Czechoslovakian mines which she has taken over. That she should have taken such early action might perhaps be understood on the ground that the son of the German Undersecretary of State, Von Weizsäcker, is attached to the Kaiser-Wilhelm-Institut in Berlin where some of the American work on uranium is now being repeated.

Roosevelt, after reading the letter, called in an aide and announced, "This requires action."

Even though this letter was delivered to the White House on October 11, 1939, relatively little action was taken in the U.S. before 1942. Early government support of the possible bomb project was limited to grants totaling some $300,000 to American universities for nuclear research projects. The program did not begin in earnest until 1942, with the formation of the *Manhattan Project*, the code name given to the vast, widespread efforts that were to lead to the successful construction of a fission bomb.

Fission may occur not only in uranium but in several other materials as well. The fission process in natural uranium (more than 99% of which is U^{238}) is best accomplished with *slow* neutrons. The trouble with this is that in the chain reaction that is a necessary part of a bomb, *fast* neutrons are produced. Further, although U^{238} undergoes fission under bombardment by fast neutrons, the probability of fission is very low and, for certain neutron speeds, there is a high probability that neutrons will be captured by the U^{238}. But the much-less-abundant uranium isotope, U^{235}, undergoes fission with either fast *or* slow neutrons. An isotope of the synthetic element plutonium, Pu^{239}, was also found to be a promising candidate for fission reactions. Thus at least two types of bomb appeared theoretically possible.

In any potential chain reaction, it is possible that neutrons produced by the fission of one nucleus will fail to be captured by other nuclei, and thus will be lost to the chain reaction. If more neutrons escape than are produced by fission, the reaction will not be self-sustaining and will cease. To prevent this from happening, it is necessary to have an amount of fissionable material large enough that more neutrons will produce fission reactions than will escape. The minimum amount of fissionable material required to initiate a chain reaction is known as the *critical mass* of that element for fission (Fig. 4.15).

In 1942 no one had a very good idea just how much material constituted a critical mass of uranium, though experiments indicated it was somewhere between 1 and 100 kilograms. Up until then, the total amount of U^{235} that had been separated from the more abundant isotope, U^{238}, was measured in *millionths* of grams. So the initial problem of obtaining enough fissionable material to construct a bomb appeared almost insurmountable.

Nonetheless, the Manhattan Project mounted an almost unbelievable effort on several fronts. Under the capable leadership of General Leslie Groves, several teams of the top scientists attacked the many problems facing them. Because so much remained to be learned in so short a time, in order to ensure success it was decided to proceed in as many directions as seemed fruitful, with the less successful projects being dropped along the way. Therefore, work was pointed toward building both uranium and plutonium bombs. There appeared to be at least three different possible methods of obtaining U^{235} and two for obtaining Pu^{239}, and efforts were made to produce fissionable materials by each of these methods.

You must remember that at this time no one had ever produced a chain reaction that was

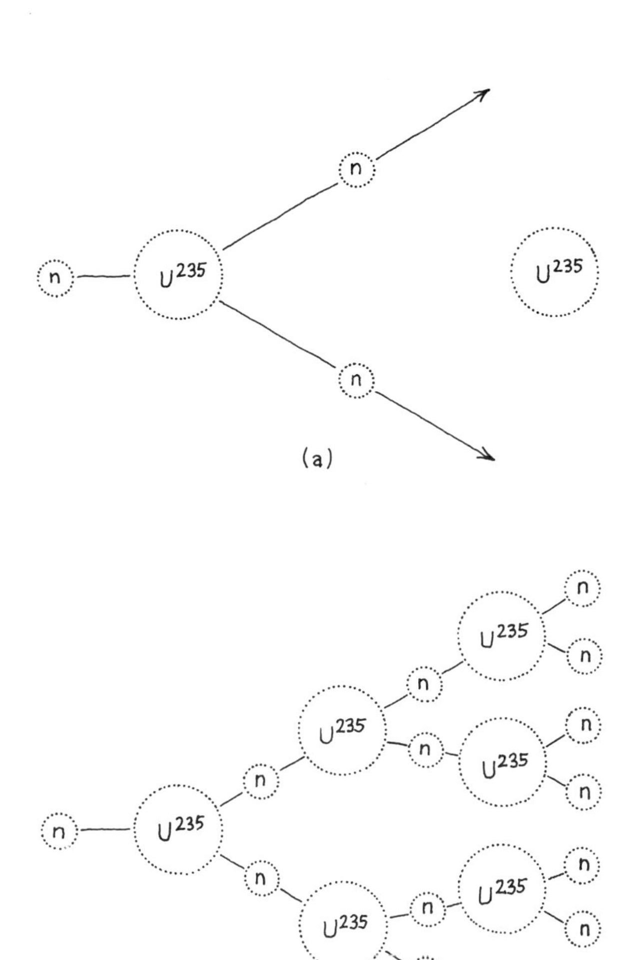

FIG. 4.15(a) Subcritical mass—neutrons escape.
(b) Critical mass—chain reaction.

self-sustaining. If the production of the bomb were actually to be shown possible, it was necessary to first demonstrate the existence of a self-sustaining nuclear chain reaction. Under the west-side stands of Stagg Stadium at the University of Chicago, on what had once been the squash court, this pioneering effort got under way. There the world's first atomic "pile" was constructed, under the direction of Enrico Fermi. Both enriched uranium and uranium oxide were used for fuel, as there simply wasn't enough pure uranium available. Neutrons were moderated by graphite blocks, which slowed them down to the low energies desired for the reaction. Graphite, or carbon, is a very common substance, but the graphite used in this pile had to be very pure; any impurities might have absorbed the neutrons produced in the reaction. Of course, one does not want to create a bomb in these circumstances, so control rods of cadmium, a material that is a very efficient absorber of slow neutrons, were placed between the sections of the pile that contained uranium. To operate the pile, the scientists withdrew the control rods one by one until the increased flux of neutrons produced a self-sustaining reaction.

On the afternoon of December 2, 1942, the Stagg Field pile "went critical"—that is, the number of neutrons produced was slightly greater than the number of neutrons lost and absorbed. The reaction was self-sustaining! The men involved were now convinced that a bomb was possible.

The success of this first uranium reactor was important for yet another reason: the operation of this type of pile was the key to the production of plutonium. Production of the synthetic element plutonium requires, first, the production of neptunium. Neptunium is obtained when U^{238} captures a neutron and emits an electron. The neptunium then beta-decays into plutonium. Though plutonium could be produced in a reactor similar to the one under Stagg Field, the *size* of a reactor capable of producing plutonium in the desired quantities was staggering. The Chicago pile operated at a power level of $\frac{1}{2}$ watt on that December afternoon, and ten days later was operated at a power level as high as 200 watts. But to produce enough plutonium to build a bomb in a reasonable length of time called for a reactor that could be operated at 1,500,000 kilowatts. To give you a basis for comparison, the Grand Coulee Dam produces 2,000,000 kilowatts.

Because plutonium is a chemical element, it can be separated from uranium or neptunium by chemical means after being produced. However, no chemical technique can be used to separate U^{235} from U^{238} because they are the same element. It was necessary to devise a physical process that would take advantage of the slight difference in mass between the two isotopes. At the start of this work, the two ideas that appeared to be most promising were gaseous diffusion and electromagnetic separation.

In the gaseous-diffusion process, a gas—uranium hexafluoride—is passed through a long series of semiporous membranes. The uranium-235 hexafluoride gas molecules can diffuse through each barrier slightly more easily than those molecules containing U^{238} and, after passing thousands of barriers, the gas at the end of the process will be enriched in U^{235} (Fig. 4.16). The electromagnetic-separation technique involves letting ions of the two isotopes traverse a magnetic field.

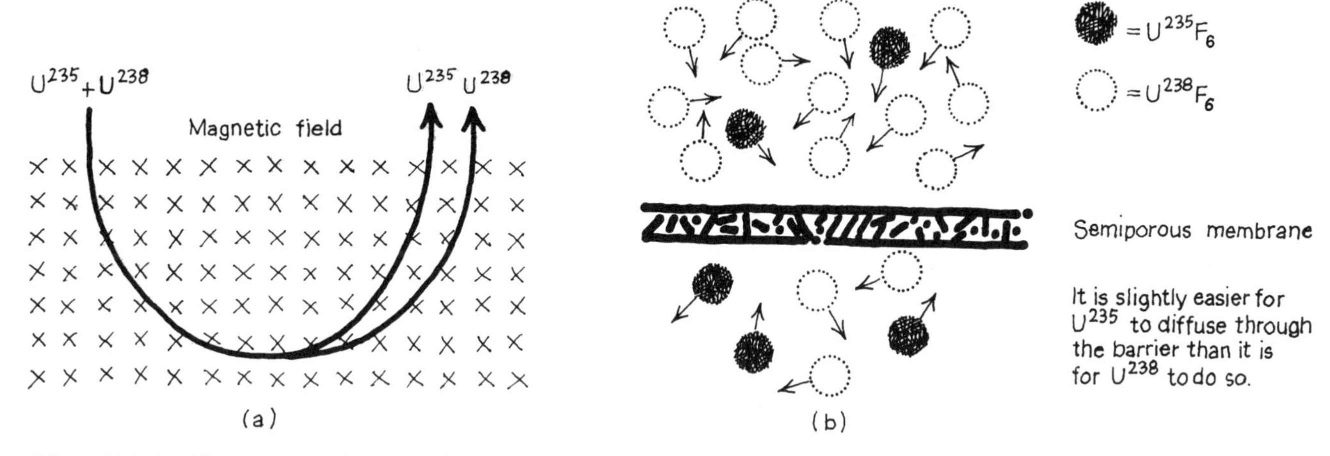

FIG. 4.16(a) Electromagnetic separation of uranium. (b) Gaseous-diffusion separation of uranium.

The slightly lighter U^{235} ions are bent in a shorter arc than those of U^{238}, thus bringing about the separation.

Both these processes are very involved ones, but, due to the press of time, the normal industrial process of building small-scale pilot plants and thoroughly testing various separation methods could not be followed. Each operation undertaken involved a tremendous gamble, yet the stakes were so high and the danger of *not* beating Germany to the production of a bomb so great that the risks were accepted.

The plutonium-separation process took place in a complex built in Hanford, Washington. This remote area was chosen not only for the security it afforded, but because the Columbia River was available for cooling purposes. The U^{235} separation was done in a new town built in the isolated hill country of eastern Tennessee, named Oak Ridge. Here a gaseous-diffusion plant, code named K-25, an electromagnetic separation facility, named Y-12, and —before production was completed—a third separation facility were located. As a result of wartime material shortages, there was not enough copper available to wind the magnet coils of the electromagnetic separator, so the Manhattan Project requisitioned 14,700 tons of silver from the U.S. Treasury. (Silver is a better conductor of electricity than copper, anyway.) The last of this nearly half a billion dollars' worth of silver was returned to the Treasury in May 1970.

Oak Ridge, Tennessee. Preparing to remove plugs from some of the 1248 fuel-channel openings in the shield of the Oak Ridge graphite reactor. Personnel stand on an elevating platform, using "push" rods to remove plugs from the reactor. (Photograph courtesy USAEC)

The bombs were to be assembled and tested by a group headed by the American physicist Robert Oppenheimer on an isolated mesa in New Mexico, working at the site of what had once been the Los Alamos Ranch School, a private academy for boys. In July of 1945 a plutonium test bomb was assembled and exploded near Alamogordo, a desert location some 200 miles from Los Alamos. No one knew how well the test bomb would work, or indeed whether it would work at all. To calibrate the instruments that were used to measure the effect of the explosion, 100 tons of TNT were detonated at the test site. When the actual bomb went off, it produced an explosive effect nearly equal to that of 20,000 tons of TNT.

Immediately after the results of the first successful bomb test, a second bomb was rushed to the Pacific to be used against Japan. On August 6, 1945, an American B-29 bomber, nicknamed the *Enola Gay*, dropped a uranium bomb on the unsuspecting Japanese city of Hiroshima, killing about 200,000 persons. Three days later, the remaining plutonium bomb was dropped on Nagasaki, killing about 74,000 persons. On August 11, Japan surrendered.

Top: "Little Boy," the first atomic bomb ever used in war, was a uranium bomb detonated over Hiroshima, Japan, on August 6, 1945. The bomb was 28 inches in diameter, 120 inches long, weighed about 9000 pounds, and killed more than 200,000 people. Bottom: "Fat Man," a plutonium bomb, was the second ever used in war. It was detonated three days later over Nagasaki, Japan. This bomb was 60 inches in diameter, 128 inches long, weighed about 10,000 pounds, and killed 74,000 people. (Photographs courtesy Los Alamos Scientific Laboratory)

Mushroom-shaped cloud created by the plutonium bomb dropped on Nagasaki.

4.7 THE GERMAN BOMB PROJECT

The great compelling drive to first complete a fission bomb had been predicated upon the Allies' fear that the first such bomb might be made in Germany. From every objective standpoint, at the beginning of the war it appeared likely that the bomb would first be completed in Germany, if such a bomb could be made at all. The advantages initially possessed by Germany included the following:

1. Virtually all the fundamental nuclear research performed before 1939 had been done in Europe, most of it in Germany.
2. During the 1930's most of the best nuclear scientists in the world were either working in Western Europe or had been trained there.
3. Germany apparently had an ample supply of raw uranium ore available from the Joachimstral mines of Czechoslovakia.
4. It seemed clear, at least from this side of the Atlantic, that the Hitler government would give every priority to the development of a weapon that promised to be as destructive as a nuclear bomb.

To this list must be added the fact that, before the end of the war, Germany developed a ballistic missile (the V-2 rocket) that might have been capable of delivering a nuclear bomb.

There is no question but that, as late as the spring of 1942, Germany could have been much further along toward producing a bomb than the United States. Under the direction of one of the world's most outstanding nuclear physicists, Werner Heisenberg, a German nuclear pile achieved a 13% multiplication of neutrons in May of 1942. This neutron multiplication was achieved nearly six months before the American group had their success with the Stagg Field reactor in Chicago. Any early German advantage was not followed up, however, and by the end of the war a self-sustaining nuclear reactor had still not yet been built in Germany.

The "Why?" behind the failure of the German war machine to produce a bomb is still a topic of debate. We can pinpoint some of the reasons involved, though their rank of relative importance may remain uncertain. It seems now that the military potential of the nuclear experiments was never brought home to those at the highest levels of the German government. The most important potential application of nuclear devices appeared to most German leaders to be the generation of electric power and most of the nuclear research was apparently directed along these lines. Early military successes apparently influenced war-effort planning, so that in 1942, at the time when a major commitment to a nuclear project would have to have been made, the prevailing attitude in many parts of the German regime was that the war would be over within six months; thus weapons systems that could not be delivered within that time were not encouraged. Finally, the intense Allied bombing, which caused the site of the bomb work to be moved twice, surely aided in rendering the project ineffective.

When one speculates on what would have happened if a German bomb had been produced that could have been delivered by V-2—or possibly longer-ranged—rockets, one may shudder to think of the outcome of the war and the makeup of our world today.

4.8 THE CONSCIENCE OF THE NUCLEAR SCIENTISTS

With the ending of the war in Europe, in May 1945, some of the scientists who had been working on the Manhattan Project felt that work on the bomb could now be abandoned. It was obvious that Japan could not possibly construct such a bomb, and the defeat of Japan was but a matter of time. Some of these men, whose consciences had permitted them to proceed in trying to develop a bomb while the fate of the free world hung in the balance, were now very unenthusiastic about continuing work on a bomb that would be used against a nation already on the verge of defeat.

There was, however, the considerable military problem of the actual invasion of the Japanese islands. Though this campaign seemed assured of success, it would probably cost a higher toll of American lives than the entire preceding part of the war. The willingness of Japanese airmen to sacrifice their lives in defense of their homeland had already been evidenced in the Kamikaze attacks at sea: Japanese pilots would deliberately crash their bomb-laden planes onto American naval vessels, sacrificing their own lives in order to inflict a heavy toll on their enemy. American leaders felt that the awesome destructiveness of the A-bomb might be used to induce an early surrender.

Some of the bomb-project scientists suggested that a demonstration of the bomb be held on an uninhabited Pacific atoll before it was actually dropped on Japan. The reasoning given was that if a team of Japanese experts could see first-hand the destruction that awaited them, they would immediately sue for peace and the bombs would not be used against their islands. However, it was argued to the contrary that such a demonstration would not necessarily be convincing nor would it accomplish the desired result, for those in the Japanese government who had not seen the demonstration would probably think it only a trick. A complicating factor in the debate was the fact that there were only two bombs in the U.S. arsenal, and months would pass before enough uranium or plutonium could become available for the production of another one. In the end, the decision to drop the bomb was made by one man, President Harry S. Truman, who willingly accepted complete responsibility for his decision.

From a military standpoint, the use of any weapon of which only two samples existed, when more could not be made for months, seemed questionable. Nevertheless, as long as the enemy did not *know* how many bombs had actually been produced, they still might be used effectively. It must also be pointed out that all previous Allied bombing operations had been supposedly directed against military targets, though civilians had been killed in those raids through the unfortunate circumstance of their being near bomb explosions. The nature of the nuclear bomb, however, was such that the primary death toll would be among the nonmilitary population in the target area, and the primary material damage would be to civilian buildings.

The results of the two bombs that were dropped on Japan are now well known. The destruction that resulted from the first bomb, at Hiroshima, was of such magnitude that Japanese government officials who read the reports, but had not actually surveyed the damage, were incredulous. At the highest levels in Japan it was by no means certain just what had happened, though there were several capable Japanese physicists who knew that only a nuclear device could have caused the destruction they had witnessed.

When the second bomb was dropped, just three days later, Japan must have been given the impression that a whole arsenal of bombs was awaiting delivery. In truth there were no more completed bombs, but the bluff worked, and the Japanese government sued for peace on August 11, 1945.

4.9 THE HYDROGEN BOMB

Following the successful development of the fission-type bomb, which used either uranium or plutonium, speculation began in some quarters about the advisability of developing an even more powerful type of bomb: a *fusion* device. The fusion bomb makes use of reactions believed to be involved in the energy-production processes of stars. In these reactions, four hydrogen atoms are combined—or fused—into a single helium atom. Just as in the case of the fission bomb, the end product of the reaction has a smaller mass than the sum of the constituent parts that went into it.

TABLE 4.2

${}_1H^1$ = 1.00782	
${}_1H^1$ = 1.00782	${}_2He^4$ = 4.00260
${}_1H^1$ = 1.00782	
${}_1H^1$ = 1.00782	
4.03128	

From looking at Table 4.2 you can see that the mass of four hydrogen atoms is greater by about three-quarters of 1% than the mass of the helium atom that is produced. In the fusion reaction, this difference in mass is converted into energy. Although the energy released from one fusion event is only about 10% of the energy released in the fission of one uranium atom, hydrogen is much more readily available than uranium or plutonium, and a larger bomb may easily be constructed.

To initiate the fusion reaction requires temperatures as hot as those in the interior of the sun. To produce these temperatures, a fission bomb is first detonated, and the resulting extremely high temperature initiates the reaction of the fusion bomb. A third stage may easily be added by surrounding the fusion-bomb material with more fissionable material to produce a fission-fusion-fission device. In this type of bomb, the energy released is about evenly divided between the fission and fusion devices.

Although it was suggested immediately following World War II that such a weapon was now possible, no all-out effort was then made in the U.S. to produce a fusion bomb. The United States had emerged from the war as the world's only nuclear power, and though it was readily believed that Russia would produce a bomb on its own, the guesses as to when this would be done varied widely. Almost no one suggested that before the decade of the 1940's was over the Soviet Union would have joined the "club" of nuclear powers.

Yet in September of 1949 an American Air Force WB-29, flying a relatively routine mission from Japan to Alaska, detected strong traces of radioactive material high in the atmosphere. The conclusion that Russia had detonated a nuclear device was inescapable.

The United States then began a hastily accelerated development program that led to a thermonuclear detonation on the Pacific atoll of Elugelab on November 1, 1952. The device that was detonated could hardly be called a deliverable weapon, for it weighed 65 tons and used the hydrogen isotope tritium in liquid form. Less than a year later, evidence was gathered that the Soviet Union had also exploded an H-bomb.

The first uranium and plutonium bombs had exploded with the destructive power of about 20,000 tons of TNT (20 kilotons). The first U.S. H-bomb was equal to about 3 million tons (3 megatons) of TNT, and the second series of bombs tested were 20-megaton-class weapons. Bombs of 100 megaton and greater size may now be constructed, though for tactical reasons bombs are made in many smaller sizes.

The testing of nuclear weapons in the atmosphere by both the U.S. and Russia continued until 1963, when a treaty (now signed by 103 nations) was adopted which banned atmospheric bomb tests. Underground weapons tests by these nations have continued since that time, however. Red China, not a signer of the atmospheric-test-ban treaty, has continued atmospheric weapons testing.

4.10 THE NUCLEAR CLUB

The United States was the first possessor of both the so-called A-bomb and the H-bomb. ("A-bomb" is a misnomer, for the energy release is due to nuclear—not atomic—reactions. Nonetheless, this popular term's wide usage is perpetuated in the very name of the U.S. Atomic Energy Commission.) The Soviet Union, a nation that also possesses a highly developed technology, was able to produce both weapons shortly after the United States did. France and, more recently, Red China have also joined the group of nations that have succeeded in detonating H-bombs.

The know-how required to develop an A-bomb is now by no means a deep, dark mystery. Almost any industrialized nation in the world which has a supply of fissionable material and a nuclear reactor, and which is willing to spend the necessary funds, can develop a fission-type bomb of its own. Among those nations that might fit into this category are, at least, Israel, Egypt, and India. This is not to suggest that these nations have built or will build bombs, but it is at least within the realm of possibility that these or other similar small to medium-sized powers might develop nuclear weapons.

Assuredly, the development of a bomb and the development of a weapons system that includes reliable delivery capability are two quite different things. Any nation planning to use a nuclear weapon would almost certainly find it necessary to first test the device. If this were not done and if, for some reason, it failed to explode, not only would the intended victim have the source of a possible political coup, but it would have gained a critical mass of fissionable material as well. Nonetheless, it is often suggested that the next use of nuclear weapons in warfare may be made by "second-rate nations fighting a two-bit war." The fear remains that the rest of the world might then become involved in a confrontation that could result in the virtual annihilation of the human race as we know it.

Recognizing just these possibilities, a group of nuclear scientists began the publication of a journal devoted to science and public affairs, *Bulletin of the Atomic Scientists,* in 1945. Prominently featured on the cover of the magazine was a clock with its hands near midnight. Following the explosion of the first Russian H-bomb in 1953, the hands were set at two minutes until midnight, but after the adoption of the atmospheric-nuclear-test-ban treaty, the hands were reset to ten minutes until twelve. The clock on the cover of this journal served as a mute reminder that we are truly living in the Age of the Bomb.

FOR MORE INFORMATION

Richard G. Hewlett and Oscar E. Anderson, Jr., *A History of the United States Atomic Energy Commission, Vol. 1: The New World, 1939–1946,* University Park, Pa.: The Pennsylvania State University Press, 1962. A scholarly record of the U.S. bomb project from a historical viewpoint.

Henry D. Smyth, *Atomic Energy for Military Purposes,* Princeton, N.J.: Princeton University Press, 1945. The first declassified information released on the U.S. bomb project. (In paperback.)

Laura Fermi, *Atoms in the Family,* Chicago: University of Chicago Press, 1954. Behind-the-scenes glimpses of scientific history in the making, as told by Enrico Fermi's wife. (In paperback.)

Robert Jungk, *Brighter than a Thousand Suns,* translated by James Cleugh, New York: Harcourt, Brace, 1953. A popularized personal history of the men who built the bomb.

Paul R. Baker, *The Atomic Bomb: the Great Decision,* New York: Holt, Rinehart and Winston, 1968. Readings on the decision to drop the bomb on Japan, and postwar world problems. (In paperback.)

David Irving, *The German Atomic Bomb,* New York: Simon and Schuster, 1967. An account of nuclear efforts in wartime Germany. (There was no "German atomic bomb.")

Peter Michelmore, *The Swift Years: the Robert Oppenheimer Story,* New York: Dodd, Mead, 1969.

Philip M. Stern, *The Oppenheimer Case: Security on Trial,* New York: Harper and Row, 1969. Two views of the controversy that surrounded the one-time director of the Los Alamos Laboratory.

QUESTIONS

1. What point of view of nature was held by those who adopted the original "atomistic" belief?
2. What has plum pudding got to do with atoms?
3. Describe Rutherford's crucial experiments on nuclear size.
4. How could elements be correctly ordered in the periodic table before the discovery of the neutron?
5. What is deuterium oxide? Tritium oxide?
6. What is the origin of the electron emitted in beta decay?
7. Based on its name alone, do you have any evidence as to when element number 94 was named?
8. What Nobel laureate in physics did experiments in a goldfish fountain?

9. If U^{235} released an average of only 1.5 neutrons per fission event, would a chain reaction be possible?
10. Who possibly made the first observations of nuclear fission?
11. Who correctly interpreted the result of Hahn's experiment?
12. What evidence was available in 1939 that Germany might be planning to build a bomb?
13. Samples of natural uranium can be made to fission. Why can't it be used for a bomb?
14. Name materials other than uranium that will undergo fission.
15. What did the Stagg Field reactor demonstrate?
16. Why did it appear possible that Germany could have developed a bomb before the U.S.?
17. Why *didn't* Germany develop the first bomb?
18. Why was there no fear that Japan was building a bomb?
19. What technical development in Germany would have made German production of a bomb particularly terrifying?
20. What is the basic difference between fission and fusion bombs?
21. What physical principle makes both fission and fusion bombs work?
22. What unexpected event speeded up work on the U.S. H-bomb?
23. Which nations have now detonated nuclear devices?

PROBLEMS

1. During World War II, the U.S. developed an A-bomb, and during the decade of the 1960's the U.S. landed a man on the moon. In each case the U.S. started behind, spent enormous amounts of money and succeeded in projects that have since become sources of controversy. Further compare and contrast these projects.

2. Is it possible that a "conspiracy of silence" among those able to do research and development in nuclear weapons would have prevented the making of a bomb? For how long? What might have been the consequences of this action?

3. If you had been president of the U.S. in August 1945, how would you have employed the A-bomb in bringing World War II to an end? Defend your reasoning.

4. Suppose that Germany had developed an A-bomb and had been able to deliver it over England with their V-2 rockets. How might this have changed the outcome of World War II? What if a German missile carrying an A-bomb could have flown the Atlantic?

5. Since it is feared that small nations who develop A-bombs may plunge the world into a nuclear holocaust, what steps might be taken as safeguards against this possibility?

6. Pu^{239} releases an average of 2.7 neutrons per fission, compared with 2.5 for U^{235}. Which of these two elements do you suppose has the smaller critical mass? Why?

7. There was some slight speculation—before the event—that the detonation of the first A-bomb would start a chain reaction that would engulf the whole earth. What arguments could you have made, beforehand, to convince someone this was not possible?

8. The uranium bomb dropped on Japan used a small cannon in the mechanism that initiated the explosion. Can you deduce how this device worked? What special features would this gun have needed?

9. What reasons can you give for the key U.S. bomb facilities in World War II—Oak Ridge, Hanford, and Los Alamos—being located where they were?

10. Suppose that an erudite postal clerk found an envelope addressed to: Professor (blank), 97,98. Where would he deliver this letter?

CHAPTER 5 NUCLEAR EFFECTS AND APPLICATIONS

5.1 NUCLEAR POWER PLANTS

The awesome destructiveness of nuclear weapons served as terrifying evidence of the vast amounts of energy that may be released by nuclear fission reactions, but the uncontrolled release of this energy in an explosion appeared far removed from any sort of beneficial use. Yet even before the first bomb had been exploded, the atomic scientists realized that the great quantities of energy derived from nuclear reactions could also be used in peaceful applications. Today this energy is being used in nuclear electric generating plants, as well as in other ways, and the secret of a bomb once destined only for destruction can now be used for the good of all mankind.

The world's first nuclear reactor was built in Chicago in the fall of 1942 as a part of the Manhattan Project. Before it would appear reasonable to continue work on a bomb, it was necessary to prove that a chain reaction was actually possible, and it was in that historic device that the first self-sustaining fission chain reaction took place. This reactor, or pile, contained four essential types of components common to all nuclear reactors:

1. Fuel
2. Moderators
3. Control rods
4. Shielding

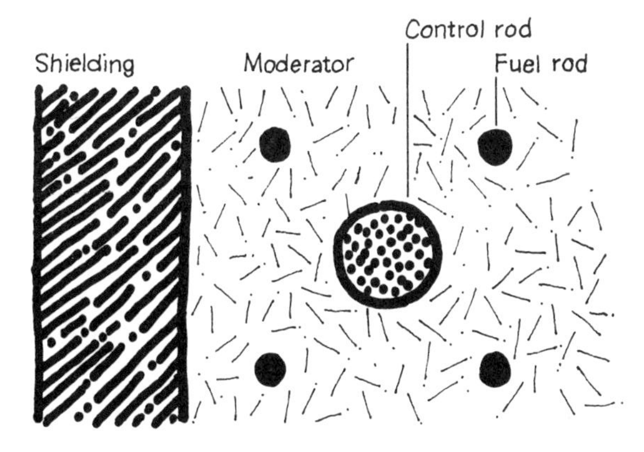

FIG. 5.1 Schematic of section of nuclear reactor.

In the Chicago reactor the fuel was uranium and uranium oxide. Graphite (carbon) was the moderator, and the control rods were made of cadmium (Fig. 5.1).

These components, along with the shielding necessary to protect the men and equipment around the reactor, are the basic parts of any nuclear reactor itself. To extract energy from the reactor a fifth component, a *heat exchanger,* is required. No provision was made for the removal of heat from the first test pile, which

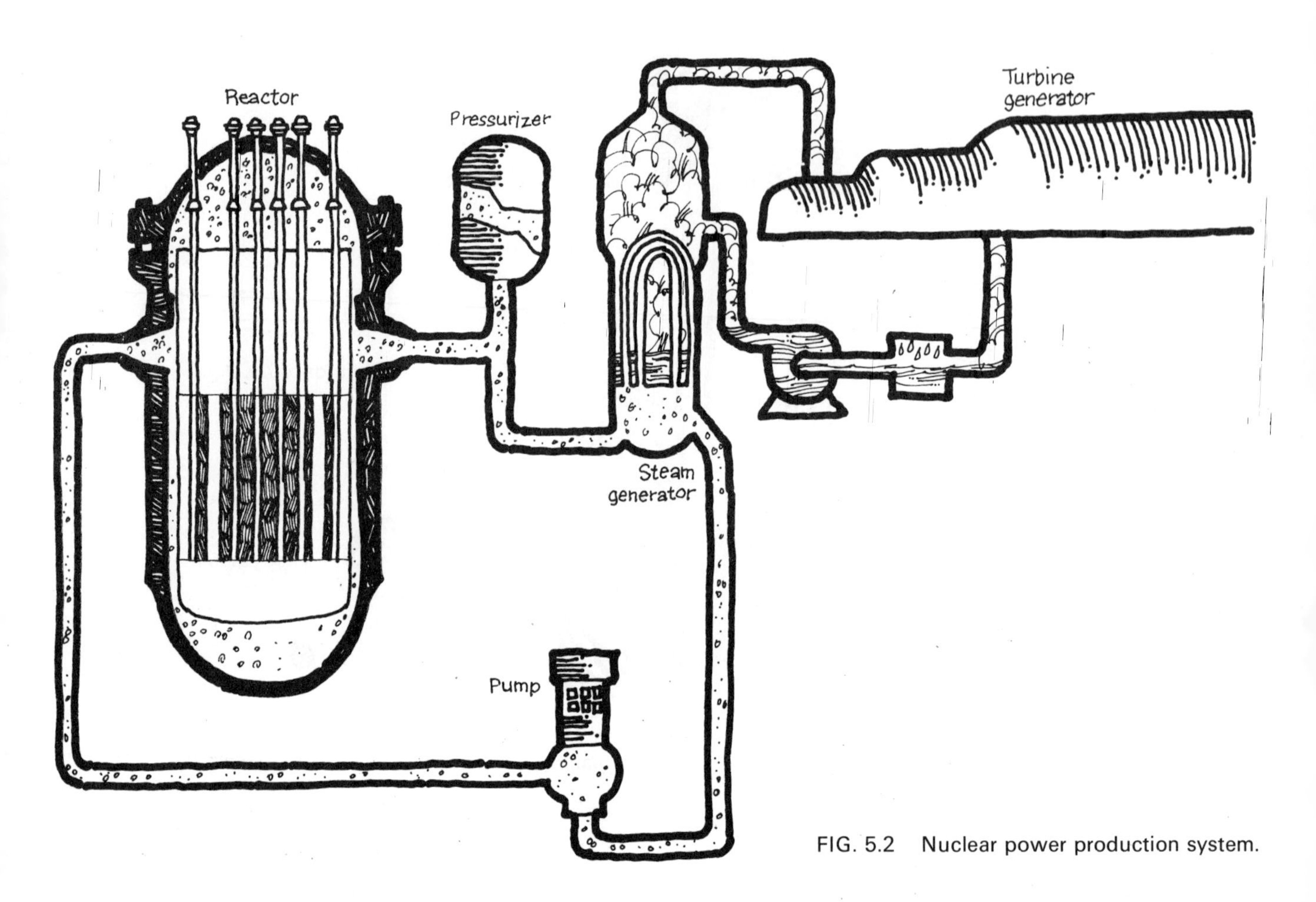

FIG. 5.2 Nuclear power production system.

severely limited the power level at which it could be operated.

A nuclear-power reactor produces energy in the form of *heat*. Conventional thermoelectric power plants burn coal or oil or gas *(fossil fuels)* to produce heat, which converts water into steam, which then turns turbines connected to electric generators. Thus—except for the way in which heat is obtained—the actual power-production process in a nuclear-generator plant may be virtually identical to that of a conventional plant (Fig. 5.2).

Care must be taken, of course, to confine the radiation released to the reactor itself. If the material that is heated directly by the pile is allowed to pass through turbines and then escape, damaging amounts of radiation may be released. For this reason, nuclear power plants employ a type of heat exchanger, in which the reactor coolant itself is always confined. Either a liquid or a gas may flow through the reactor core and transfer heat to the working fluid that drives the turbines. It is also possible to convert heat directly into electrical energy without going through the intermediate steps of producing steam and turning turbines and generators, and we shall describe the direct conversion of heat into electricity in Chapter 8, during our discussion of thermoelectric devices.

At this point you may be wondering why there has been so much excitement over nuclear power generation if it is so similar to conventional power production once the heat energy has been produced. The answer lies in the different form in which the fuel is found and used and in important operational differences, involving size and costs, between conventional and nuclear power production.

All power plants are simply energy-conversion devices. Energy in the form of falling water may be used to turn generators and produce electricity, energy in the form of heat from burning fuel may turn water into steam which rotates turbines, etc. In any case, to effectively produce electrical power, one must have available stored energy in a form suitable for this conversion. Energy may be stored in the waters of a reservoir behind a dam, and released as the water falls to a lower level, or in a fossil fuel, which releases heat when burned. Unfortunately, in many areas of the world none of these energy sources are readily available. In such areas the production of electrical power usually means transporting fuel over great distances, and these fuel costs may make the power produced too expensive for the extensive development of that area.

The amount of fuel needed for a nuclear power plant is limited to literally a few pounds of uranium or other fissionable material. Not only is the amount of fuel small enough to eliminate most transportation problems, but the fuel lasts long enough to keep these transportation costs negligible. In a type of reactor known as a *breeder reactor*, the reactor not only does not use up the fuel with which it is supplied, but actually *converts* nonfissionable material into fissionable material, providing more fuel than is actually used! This process might be compared with the delightful situation you would experience if you were to fill your car's gas tank half full, drive several hundred miles without adding more fuel, and find that at the end of your trip the tank contained more fuel than you had put in it originally. Breeder reactors may even be used to produce salable fuel at the same time that they are producing power.

Nuclear power plants are now in operation and under construction in many parts of this country, as well as other parts of the world. At the moment the cost of a nuclear power plant is comparable to that of a conventional plant but, even in areas in which alternative power sources are available, it is anticipated that the operating costs of nuclear power plants may make them economically more feasible than conventional plants. In those areas without the resources for economical power production by conventional means, nuclear power plants may make possible the industrialization of areas or entire nations that are presently underdeveloped.

Nuclear power plants are particularly well suited for use in underseas vessels. Conventional underseas craft operate diesel-powered (air-breathing) engines, while on the surface, to charge batteries which are then used for underwater propulsion. Because the primary engines

Commercial nuclear power reactor fuel assembly. This unit contains 204 fuel rods with 41,412 uranium fuel pellets. (Photograph courtesy of Westinghouse Electric Co.)

cannot be run underwater to recharge these batteries, the undersea range of conventional submarines is limited to the distance the craft will operate on batteries. The underwater performance of the vessel depends on the length of its underwater stay: the higher the power level used for operation, the shorter the underseas duration. A conventional submarine that must periodically resurface is highly vulnerable to detection.

Nuclear power plants are not air-breathing; thus a nuclear-powered submarine may remain submerged indefinitely. In addition, its performance while submerged does not adversely affect the length of time submerged operation is possible. The U.S. has built a number of nuclear-powered submarines, some of which have remained submerged for periods of time longer than 6 months. This class of submarine is also designed to be able to launch Polaris-type missiles from under the sea.

Other applications of nuclear-powered devices have been found in space operations. Nuclear generators have been used to provide electrical power for the seismometers left on the moon by the Apollo missions. Nuclear-reactor propulsion units are now being developed that will propel the first manned missions to Mars and beyond.

Among the greatest *dis*advantages of nuclear power plants is the fact that shielding must be employed to protect men and materials from radiation. We shall discuss some aspects of the health problems associated with radiation in Section 5.5. Although it is possible to provide sufficient shielding to minimize radiation hazards to workers in nuclear power plants, the shielding needed adds to the size and weight of the power plant. For this reason there are applications for which nuclear power plants may not be suitable. A nuclear-powered airplane would have to be of enormous dimensions and weight to include sufficient shielding for humans carried on board. Shielding would be less of a problem for a nuclear power plant used in deep-space travel, for the power plant and the manned quarters could easily be located a considerable distance apart, with shielding necessary only between the power plant and the passengers.

Not only men but materials also may be damaged by absorption of radiation. Flexible lines and couplings become embrittled after long-term absorption of radiation. Other materials may change their characteristics and suffer failure or fatigue. Maintenance schedules at nuclear plants call for periodic replacement of those components particularly susceptible to damage.

Sufficient shielding can be made available so that radiation hazards are minimal, even for persons working at the site of the reactor itself. The precise effects on humans of very low levels of radiation are still a topic for research, and tentative conclusions are widely divergent, but although the hazards to health from low levels of radiation are debated, it is quite certain that the chemical pollutants released into the air by the burning of conventional fuels *do* contribute to sickness and death. So far as the possibility of explosion is concerned, the type of reactor used in power production does not require fast-fission materials such as those employed in bombs, and the potential for a nuclear explosion at a power plant is vanishingly small.

The question of thermal pollution by nuclear generators has also been raised. A nuclear power plant—like any other thermal energy-conversion device—requires a coolant such as air or water. The release of heat to the surroundings is common to both nuclear and conventional power plants, although nuclear-powered operations generally have greater energy-production capacities and produce greater thermal loads.

Since the fusion bomb offers so much greater energy release than the fission bomb, why are there no fusion power plants planned or built? The answer is simply that to date no one has devised a satisfactory method of confining the fusion process. Remember that the fusion process takes place at temperatures equivalent to those of a uranium-type bomb. Given these temperatures, it is obvious that no material can confine the reaction, so the most promising means yet devised is the containment of the fusion process in a magnetic field—a so-called *magnetic bottle*. Significant progress toward the production of a confined fusion reaction has recently been reported by a team of Russian scientists using a family of machines known as *Tokomaks*. The intense interest shown by several nations in the development of fusion reactors is due not only to the vast amounts of power they can release, but because—unlike fission reactors—fusion reactors will not produce large quantities of radioactive wastes, with attendant disposal problems.

The social and economic progress that could result from delivering large-scale sources of inexpensive power to the presently underdeveloped areas of the world is staggering to the imagination. Supplied with an abundance of inexpensive electric power, these now-disadvantaged regions could make rapid strides toward taking their rightful places among the nations of the world.

5.2 RADIOACTIVITY

Radioactivity is a natural property for many materials, while in others it may be artificially induced. The fact that a material is radioactive means that it spontaneously emits radiation. This radiation may consist of particles such as helium nuclei—*alpha particles*—or electrons—*beta particles*—or energetic electromagnetic radiation known as *gamma rays*. In any sample of radioactive material the level of activity declines with the passing of time. The emission of a particle by an atom in a radiation process is a random sort of thing, and one cannot predict when any given atom in a sample will emit radiation. However, in a large sample, the time required for the level of activity to diminish to half its original value can be measured quite accurately. This time is defined as the *half-life* of the sample. Once the half-life for any given material has been measured, the past and future levels of activity for that sample may be calculated.

Suppose that you had a material—call it material *A*—with a half-life of one hour (Fig. 5.3). If you started with a sample containing 1000 atoms of material *A*, after 1 hour, 500 of them would have decayed into something else—let's call this material *B*. After another hour, half this remaining sample would have decayed. Now 750 of the atoms would be *B* and only 250 atoms of *A* would be left. After yet another hour, another half of *A*—125 atoms this time—would have decayed into *B*. Now if you were to find the sample with 875 atoms of *B*, and 125 of *A* remaining, knowing the half-life of material *A* you could calculate when the sample consisted only of material *A*. (Naturally you would have to assume how much *B* material, if any, had been contained in the sample at time zero.)

This is precisely what scientists do in determining just how *old* things are that were once alive. The normal carbon isotope is C^{12}, but small amounts of a radioactive form, C^{14}, are produced by cosmic radiation. The specific reaction involves the capture of a neutron by an atom of atmospheric nitrogen, converting the normal N^{14} into N^{15}. After the N^{15} emits a proton, C^{14} is left. Carbon 14 *is* radioactive, having a half-life of 5600 years, and decays—by the emission of an electron—into N^{14}. Some of this radioactive carbon is in the carbon dioxide molecules ingested by every living plant and animal. Once the plant or animal is dead, the intake of C^{14} ceases. Thus measuring the ratio of C^{14} to C^{12} in any object that was once alive enables us to estimate its age. This method appears to give excellent

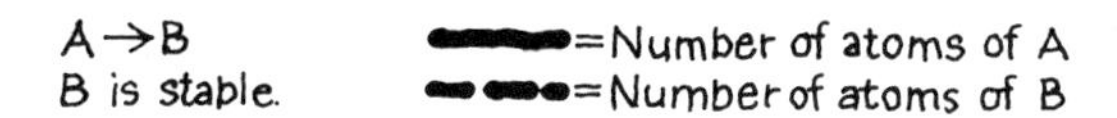

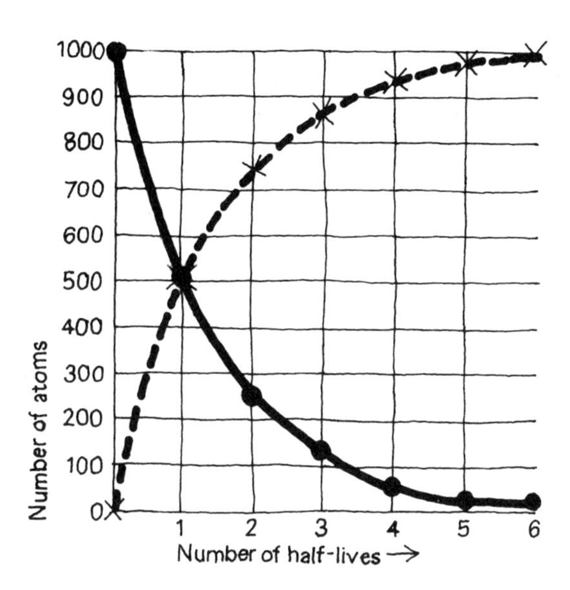

FIG. 5.3 Radioactive decay.

agreement for time periods up to as long as 25,000 years ago.

Using other radioactive elements and their decay schemes makes it possible for us to measure other time spans. From the uranium decay scheme, we can estimate that the age of the earth is about 4.5 billion years.

It is often useful to artificially induce radioactivity in a material. By "tagging" materials (i.e., making them radioactive), one can quite easily solve various tracing and detection problems. For example, one may wish to know the distribution of a certain element in the human body. A way of doing this is to introduce some of the element in a very mildly radioactive isotopic form, and then scan the body with radiation detectors. By this method one may determine how even minute traces of any specific element are distributed in the body. Comparing the distribution of a given element in a normal body and in one suffering from some malady has greatly helped to advance our knowledge of the treatment of bodily malfunctions.

In many industrial operations, the addition of traces of a radioactive substance simplifies marking or tagging products. Years ago locating buried pipelines was a laborious and time-consuming task. Now a material that is very slightly radioactive may be allowed to flow through a pipeline while a worker above the ground traces its exact path with the aid of a radiation detector.

5.3 NEUTRON ACTIVATION ANALYSIS

Often one wishes to determine just which elements are present in a sample of material. To make this determination chemically, a portion of the material to be analyzed must be used for testing. However, frequently it is undesirable to destroy or damage even a small portion of the sample—as would be the case for an oil painting or a vase—or the sample may be inaccessible, as in the case of underground mineral deposits. An alternative method of detecting small traces of elements is to make them radioactive and observe their decay patterns. Each nucleus has unique, identifiable, decay schemes in terms of the particles emitted, their half-lives, and the energies of the emitted particles. Thus each element may be identified from an analysis of these characteristics. For example, when the stable chlorine isotope C^{37} is bombarded by neutrons, it becomes C^{38}, which may decay by the emission of an electron into Ar^{38}, with a decay half-life of 37 minutes. The emitted electron is accompanied by a group of gamma rays, each having a distinct energy. No other nuclear decay scheme involves this particle with these energies and this particular half-life.

This technique we are describing is known as *neutron activation analysis.* An efficient method of inducing artificial radioactivity in an element is to bombard it with neutrons. The decay scheme of the now-radioactive material is then carefully analyzed to determine what elements were actually bombarded.

One of the first applications of neutron activation analysis was in the petroleum industry. Deposits of oil or other materials may be located below the ground by this technique. Geologists have been able both to locate mineral ores and to estimate the size of the deposits by neutron activation analysis.

In medical applications, trace elements in the body may be detected by first bombarding an area with neutrons and then measuring the radioactive decay schemes of the elements. Unlike the process we have previously described, in which radioactive trace elements are introduced into a living organism and their distribution determined, neutron activation analysis makes possible the detection of traces of whatever elements are already present in the organism. Another advantage of neutron activation analysis over radioactive tracer techniques is that the subject normally receives a smaller dose of radiation during the measurements.

One of the most fascinating applications of this technique, carried out in 1961, involved measurements on locks of Napoleon's hair. It had long been suspected that the French emperor did not die on St. Helena of natural causes. After Napoleon's death his head was shaved and locks of his hair were sold as souvenirs. Well-documented samples of this hair, when subjected to neutron activation analysis, revealed that his death was indeed unnatural: he was poisoned by arsenic. Evidence for this conclusion came from the finding of a concentration of arsenic in Napoleon's hair many times that found in a healthy in-

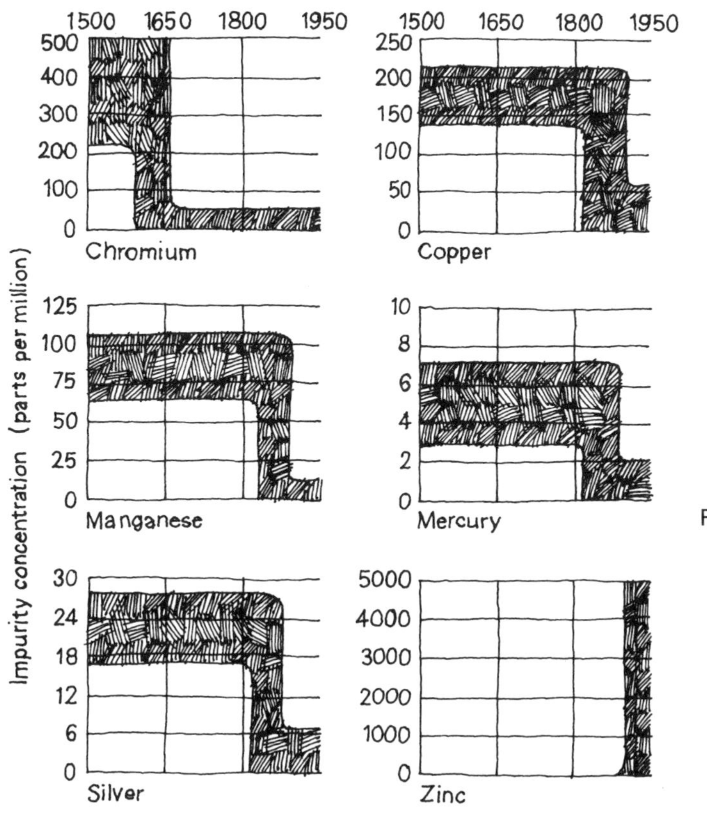

FIG. 5.4 Impurities in lead.

dividual. Further, the measurements made it possible to determine the concentration of arsenic in the various segments of this hair. Assuming a normal growth rate of his hair, it was found that the arsenic was apparently not given to him on a daily basis, but sporadically, with those days when he was—historically—reported sickest being the days of highest dosage.

Yet another application of the neutron activation analysis technique has been in testing the authenticity of paintings. The fractional concentrations of impurities in the lead pigments used in oil paintings are known to have differed for the various periods of manufacture of the pigments, since the metallurgical techniques by which lead is refined have changed (Fig. 5.4). For example, lead refined before 1650 contains chromium impurities, but since that time the traces of chromium are almost completely absent. Lead samples refined before 1850 include large traces of copper, mercury, silver, and manganese, but little zinc or antimony. Lead refined since 1950 contains little mercury or manganese, but easily detectable traces of zinc and antimony. Not only does information obtained in this way indicate the period in which a painting was done, but comparisons with other known works of an artist may provide conclusive evidence about a painting's authenticity.

5.4 IMAGING WITH NEUTRONS

Although the use of x-rays to examine materials opaque to visible light has been common for a number of years, there are several drawbacks to this technique. X-rays interact with the electrons of the atoms in the material being examined. Thus the higher the atomic number (i.e., the more electrons in the material) the more opaque the material to x-rays. For this reason, a material such as lead is very opaque to x-rays and may shield any object located behind it, while lighter materials are so transparent to x-rays that most of their detail is washed out on x-ray photographs.

However, using neutrons to obtain visible images holds great possibilities. Neutrons are scattered from the nuclei of the material being examined; thus the atomic number of the material has little effect on the scattering. Neutron-imaging may become much more widely used in coming years, as lower-cost, high-intensity neutron sources become available. Until recently, good sources of neutrons (other than reactors) were almost nonexistent. Now synthetic elements, such as Cf^{252}, and combinations of elements—such as americium-beryllium-curium—promise portable, energetic neutron sources at modest cost. Using neutron-imaging, one may obtain an image of materials that would be shielded to x-rays, and reveal details that would be impossible to see with x-rays. Although an x-ray image of an object located inside a thick-walled lead box would simply show an opaque box, neutron-imaging could also reveal the object inside the box.

5.5 RADIATION HAZARDS

Every person is exposed to radiation of various types each day that he lives. The sources of this radiation include cosmic rays, radioactive materials within the earth, naturally or artificially radioactive materials which one comes near, x-rays, fluoroscopes, and the fallout from nuclear weapons tests. Much of this radiation passes through your body without interacting with anything; this is radiation you don't need to worry about. But some of this radiation does interact with the material of which you are made, and produces changes in body material. It is this radiation we do need to be concerned about.

The changes produced in living material by radiation are not completely understood. Naturally radioactive materials emit three types of radiation, which were once called alpha, beta, and gamma rays. Alpha rays, or particles, are the nuclei of helium atoms, beta rays are electrons, and gamma rays are electromagnetic radiation of high energy. When any one of these types of radiation interacts with biological material, it loses energy to the material. One way in which energy may be lost is in the production of ion pairs. These ion pairs may trigger or speed up certain chemical reactions

TABLE 5.1

Source	RBE
x-rays	1
Thermal neutrons	2–5
Fast neutrons	10
Alpha particles or heavy ions	20

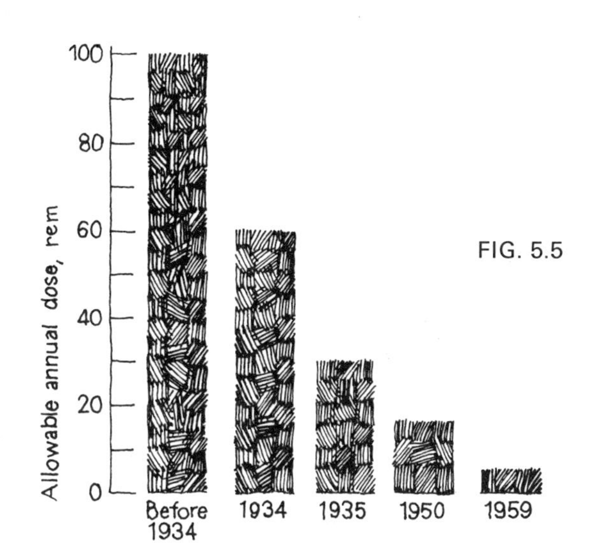

FIG. 5.5

which, in turn, may sicken, permanently damage, or even kill the person involved.

In attempts to quantify radiation dosages, scientists have formulated a number of different radiation units. Because we shall not need to use them all, we shall here be concerned only with the *rad,* defined as the amount of dosage which causes 100 ergs of energy to be absorbed by 1 gram of biological material, and the *rem*, the dosage in rads times the Relative Biological Effectiveness (RBE) factor of the radiation.* The RBE factor takes into consideration the fact that not all types of radiation interact with biological material with the same effectiveness. A list of some RBE factors is given in Table 5.1.

In discussing radiation dosage, we must consider not only the amount of radiation absorbed, but its type, distribution over the body, and even the rate at which the dosage is received. Graphic evidence of the ignorance once associated with allowable radiation dosages is presented in Fig. 5.5, which shows the recommended allowed radiation dosages prescribed by the International Commission of Radiation Protection (ICRP). Note that the allowable dose for persons working in radiation-connected industries has been reduced by a factor of 20 since 1934. This reduction is mute evidence of people's increasing knowledge of the health hazards to human beings which may result from radiation. It is usually considered that the tolerable rates of industrial

*The *rad* differs only slightly from the older unit, the *roentgen* (named after the German experimenter Wilhelm Roentgen), but is a more widely applicable and useful unit for dealing with dosages in biological materials. The *rem* is an abbreviation of "roentgen equivalent mammal (or man)."

doses are 10 times those that may be received by the entire population.

Damage to an individual exposed to radiation consists of two types of change: *somatic* and *genetic*. Somatic damage involves the individual himself, while genetic damage will affect his unborn children. Many examples of somatic damage are now known to exist in persons who were exposed to high radiation levels over long periods of time. Anywhere uranium is found there is also present an odorless, colorless, inert gas—radon—which is produced in the radioactive decay chain of uranium. Because radon is inert, conventional chemical filtering processes cannot remove it from the air a person breathes. It was found that every worker who worked in the Joachimstral uranium mines in Czechoslovakia for as long as 10 years—and lived at least 10 years afterward—died of lung cancer. It may be of interest to note that Pierre and Marie Curie, French physicists who did pioneer work with radioactive materials, were forced to do some of their early experimentation in a drafty shed, due to lack of funds. Those drafts kept the radon gas blown away and probably prolonged their lives.

The early history of the use of radioactive materials in industry abounds with examples of dangerous practices and lack of concern with radiation safety. Before World War II, some workers in the watchmaking industry painted numbers onto the dials of watches using a radium compound that would glow in the dark. Most of these workers were prone to point the tips of their brushes in their mouths. Virtually all of them later developed cancerous lesions about the mouth.

Very few studies have been possible under controlled conditions on people who were exposed to very large doses of radiation. One of the first such exposures occurred during the project leading to the construction of the first A-bombs at Los Alamos during World War II. Some of the physicists involved treated fissionable uranium with little respect. If two samples of subcritical masses of uranium are brought into contact, there is a small fission reaction between them, but this reaction separates the samples before a chain reaction can become self-sustaining. At Los Alamos, this practice became a trick known as "twisting the dragon's tail." Among those who indulged themselves in this game was a physicist, Louis B. Slotin. One day the samples he used were either larger than usual or else they came closer together, for there was a blinding flash of blue light, with the momentary joining of the materials. Slotin and his colleagues knew that a great deal of radiation—perhaps a lethal dose—had been released. Ever the scientist, Slotin refused to seek medical attention until he had carefully recorded on a blackboard the names and positions of every person in the room so that accurate measurements could be made of the relative amounts of radiation each had received. Slotin became ill shortly thereafter, and died, the first casualty of the A-bomb.

Most of the Hiroshima and Nagasaki victims were killed by the blast and the fireball produced by the bombs. Thousands of others, however, survived the blast, but received massive doses of radiation. From their experiences, plus limited information obtained in later industrial accidents, the following data have been drawn. Short-term exposures of up to 25 rems normally produce no clinical effects. With doses of between 25 and 100 rems there are slight blood changes. After one receives 100 to 200 rems, there is up to 50% incidence of vomiting within 3 hours, followed by fatigue, loss of appetite, and moderate blood changes. The prognosis for recovery is good to excellent, with recovery in a few weeks.

The receipt of between 200 and 600 rems can be fatal. Those exposed to more than 300 rems will vomit within 2 hours; there are severe blood changes, with hemorrhaging and infection and loss of hair to follow within 2 weeks. There is a recovery probability of between 20% and 100%, with recovery times lasting from a month to a year. Doses between 600 and 1000 rems produce vomiting within an hour of exposure, severe blood changes, hemorrhaging, infection, and loss of hair. Of those so exposed, 80% to 100% die within two months, and survivors may expect a long period of convalescence. Above 1000 rems the fatality rate is 100%, with circulatory collapse, respiratory failure, and brain edema producing death. The lethal dose for 50% of an exposed population is 400 rems.

Even those who do recover from large radiation dosages may suffer long-term effects of this exposure. Ten years later, 100,000 survivors of the Hiroshima and Nagasaki blasts had a death rate that was annually in excess of 11% to 12% above the death rate of the general population in the same age groups, with half these deaths being from leukemia.

For most persons, except for those who are involved in industrial accidents or are near an exploding nuclear weapon, the possibility of gross pathological damage due to radiation is slight. However, the long-term effects of an entire population's being exposed to radiation must be seriously considered.

Every person on earth is continually exposed to a certain amount of radiation from natural sources. These sources include cosmic rays, which arrive at the earth from outer space, and naturally radioactive materials on the earth. The dosage from this natural background radiation varies widely from place to place, but it is at least 0.1 rem per year, or about 7 rems during an average lifetime. The radiation dose from cosmic radiation varies with altitude and latitude: the higher either becomes, the greater the dosage. At 15,000 feet the cosmic ray dosage is five times as great as at sea level. Other than walking about in a lead suit or living in a deep mine, there isn't much that one can do about this natural dosage. However, one can also receive radiation from other sources, such as fallout from atmospheric bomb tests, color TV sets, medical and dental x-rays, and other man-made services.

Workers in radiation-connected industries are now allowed (by law) a dose of 5 rems per year or 100 mr per week. (The mr is one-thousandth of a rem.) These persons will probably always constitute only a small percentage of the population. Effects from high background levels of radiation on *entire* populations are a much more serious factor. Estimates of what is "allowable" may be made only in terms of entire populations. For example, the ICRP notes:

Briefly, the suggested limit for the allowable genetic dose (of radiation) was arrived at in the following manner: Estimates made by different national and international scientific bodies indicate that a per capita *gonad dose of 6–10 rems accumulated from conception to age 30 from all man-made sources would impose a considerable burden on society due to genetic damage, but that this additional burden may be regarded as tolerable and justifiable in view of the benefits that may be expected to accrue from an expansion of the practical applications of "atomic energy."*

Interpretation of "considerable burden on society due to genetic damage" is left to the reader.

There is now virtually inescapable evidence that there is no threshold below which radiation exposure is incapable of producing both somatic and genetic damage in humans. Recent work of Ernest J. Sternglass has pointed to an apparent correlation between excess infant mortality rates and the presence of increased atmospheric radiation levels. He found that there was an anomalous increase in infant mortality rates in a narrow band of states in the U.S.—Texas, Arkansas, Louisiana, Mississippi, Alabama, and the Carolinas—in the period just after the detonation of the world's first A-bomb in New Mexico in 1945. This increase in infant mortality rates occurred in the direct path of the fallout from that bomb test, and did not occur in neighboring states unaffected by the fallout. Similar increases in infant mortality rates were noted in the parts of the U.S. affected by the fallout from the Nevada tests of the early 1950's.

The discovery of this relationship between these excess numbers of infant deaths and nuclear blast fallout was made only after a careful study of possible correlations of effects, using a large computer. The increase in infant mortality rates, the increase in incidence of leukemia, and the rise in strontium-90 levels in milk were all perfectly correlated on a year-by-year basis. (Strontium-90 is a relatively abundant fallout product.) Because strontium is chemically very similar to calcium, it is readily absorbed by the human body and may be stored in place of calcium. The half-life of strontium-90 is 28 years; thus its rate of activity remains relatively constant throughout the human life span.

One of the most disturbing aspects that these studies have uncovered is that the infant deaths correlated to increased levels of background radioactivity did not at first seem related to fallout at all. The infants died of a variety of well-known infant diseases; they were simply less resistant to these diseases than normal babies. A second observation was that many of the babies affected by fallout were much smaller than normal, the so-called "small-baby syndrome" that has been noted with increasing frequency in the U.S. throughout the past few years. Effects of the fallout from nuclear test blasts now appear to be responsible for 1% more infant deaths than would have otherwise occurred in the United States. This excess infant mortality remains despite the fact that there have been no U.S. atmospheric bomb tests since 1963, and the effect may continue for another generation.

Critics of Sternglass' work have questioned whether increased radiation levels in the atmosphere are solely responsible for the effects he cites, and have argued that correlations also exist for other environmental factors. Questions also exist about clear laboratory demonstrations of the effects of very small radiation dosages. To the questioners, the extrapolation of large-dose data to predict small-dose results, or the application of experimental findings involving animals to human populations, is not conclusive evidence.

Other radiation sources have nothing to do with bomb explosions. X-ray machines may produce as much as 1 rem per exposure. Before World War II, when little attention was paid to radiation protection, x-ray machines were sometimes used in altogether too careless a fashion, even being employed in treatment of acne and other skin disorders. Now much more stringent standards have been imposed, and, thanks to the development of much more sensitive films, exposure times for x-ray photographs have been reduced by as much as a factor of 1000. The use of fluoroscopes, which once were placed in stores as aids in fitting shoes, has been outlawed in most states.

Even color TV sets produce low-energy x-rays. This radiation is particularly dangerous to biological material, since the absorption of x-radiation by the body is at its maximum in this particular energy range. The amount of radiation received by persons viewing a color TV from distances greater than 8 or 10 feet is slight, but a child who sits directly in front of a color set for extended periods of time may face a distinct health hazard.

FOR MORE INFORMATION

Ralph E. Lapp and Howard L. Anderson, *Nuclear Radiation Physics,* Englewood Cliffs, N.J.: Prentice-Hall, 1964.

D. J. Rees, *Health Physics,* London: Butterworth, 1967.

Samuel Glasstone (editor), *The Effects of Nuclear Weapons,* Washington, D.C.: U.S. Government Printing Office, 1964.

Willard F. Libby, *Radiocarbon Dating,* Chicago: University of Chicago Press, 1955.

William A. Higinbotham, "Nuclear Safeguards," *Physics Today,* November 1969, page 40.

Werner H. Wahl and Henry H. Kramer, "Neutron Activation Analysis," *Scientific American,* April 1967, page 68.

Ernest J. Sternglass, "Infant Mortality and Nuclear Tests," *Bulletin of the Atomic Scientists,* April 1969, page 19. (Further articles in the same journal, December 1969, page 29; May 1970, page 41.)

Ernest J. Sternglass, "The Death of All Children," *Esquire,* September 1969, page 1a.

Daniel Merriman, "The Calefaction of a River," *Scientific American,* May 1970, page 42. A preliminary report on the effects of the heating of the Connecticut River by a nuclear power plant.

QUESTIONS

1. What are the essential elements of a nuclear reactor? What does each do?
2. In what form is energy obtained from a nuclear reactor?
3. What similarities exist between conventional (fossil-fuel) and nuclear generating plants?
4. What features unique to nuclear power plants can make it possible to use them in locations in which conventional plants are not feasible?
5. What disadvantages are associated with nuclear power plants?
6. What is the unusual operational bonus from a breeder reactor?
7. What particular advantages do nuclear power plants have for use in submarines?
8. What is the current status of nuclear-fusion power plants?

9. What problems are faced in building nuclear-fusion power plants that are not encountered in nuclear-fission power plants?
10. What does it mean to say that a given radioactive material has a half-life of one hour?
11. How many half-lives must pass before all the radioactive material in a given sample has decayed?
12. You are given 100,001 atoms of a radioactive element with a half-life of five minutes. After five minutes have elapsed, how many atoms of that element do you expect to have left?
13. What properties of nuclei make possible their unique identification by neutron activation analysis?
14. Describe a famous case of poisoning that was uncovered by neutron activation analysis.
15. What does neutron activation analysis have to do with art forgery?
16. Why is imaging with neutrons more desirable for certain applications than x-rays?
17. Why has neutron-imaging not been as widely used as x-ray photography?
18. If no nuclear bombs had ever been exploded, what minimum amount of radiation would a person receive during his lifetime?
19. Other than bomb fallout, what can increase this minimum dosage received?
20. What is the significance of Relative Biological Effectiveness?
21. Describe two types of radiation damage to humans.
22. What is the history of standards set for allowable radiation dosages?
23. Relate some of the "horror stories" of radiation damage to workers before radiation hazards were well understood.
24. Describe the effects of whole-body doses of 50, 200, 600, and 1000 rems.
25. What may radiation have to do with excess infant mortality rates?
26. Why is strontium-90 a villain in radiation damage?
27. What radiation hazard is presented by color TV sets?

PROBLEMS

1. Recently a manufacturer of nuclear power facilities ran an ad that included this statement: "Johnny had three truckloads of plutonium. He used three of them to light New York for one year. How much plutonium did Johnny have left? Answer: four truckloads." Explain.

2. If exposure to radiation is so bad, why is radiation therapy used for cancer victims?

3. It has been suggested that, if a widespread nuclear war were to take place, the survivors might be forced to alter their laws and permit euthanasia (mercy killing) afterward. Why might such a step be necessary? Attack or defend its justification.

4. It has often been said in the West that Red China has no fear of a nuclear war, since it would only help solve her population problem. Is this a reasonable hypothesis? Why (or why not)?

5. The use of hidden x-ray or neutron-imaging cameras has been suggested as a sure method of foiling airline hijackers. Comment on the use of such a scheme.

6. Virtually every section of the country needs more electric power. Conventional power plants pollute the air, nuclear power plants heat up the water, and nobody wants to live near a nuclear reactor anyway. What possible solutions can you suggest as a way out of this dilemma? (Even a stable population needs more electric power to improve its quality of life.)

7. Nuclear power plants are being used in submarines, and they will be used to power space vehicles, but so far there are no immediate plans to use them in airplanes. Why?

8. A day when every home would contain its own nuclear power plant has been foreseen. Consider the practicality of this idea as compared with the current concept of central power-generating facilities.

9. Proposals have been made to employ nuclear explosions in large-scale construction projects. One plan would employ bombs to create a new Panama Canal. Rumors also exist that the U.S.S.R. may use nuclear bombs to create a vast man-made sea in Siberia. Describe possible consequences of these projects.

10. Allowed radiation doses for workers in nuclear-connected industries are higher than those thought safe for the general population. How is this difference in standards justified? Can you suggest a long-range solution to this problem?

CHAPTER 6 THE WORLD OF QUANTUM MECHANICS

6.1 INTRODUCING THE QUANTUM

So far in our discussion of atoms and atomic-sized particles, we have treated them as if they behaved very much like things behave in the world of our experiences. We visualized electrons and protons as billiard-ball-like objects and explained the fission process in terms appropriate to the behavior of a drop of water. This was a very natural and logical thing to do, in view of the universe willed us by Newtonian physics. The great triumph of Newton's laws that allowed one to explain the motions of planets by the same laws used to explain the motion of falling bodies on our earth seemed to suggest that atoms and molecules might also be treated by these laws. Indeed, there was something quite satisfying about the idea that it is possible to think of planets orbiting the sun and electrons orbiting the atomic nucleus in almost the same way.

Isaac Newton.

Though the physicists of the latter part of the nineteenth century did not know about nuclear fission or electron orbits, they had nonetheless reached a stage at which it seemed that most of the real mysteries of nature had been solved, and only a little tidying up around the edges remained. Even the great American physicist A. A. Michelson was once quoted as having said that the task of physics had been reduced to making measurements accurate to the sixth decimal place. It was known that a few difficulties remained to be cleared up before all observed phenomena could be explained, but virtually everyone was confident that with a little more effort these problems could be eventually understood within the context of nineteenth-century physics.

Yet these problems were not easily understood, and from the explanations of those very "difficulties," investigators discovered far more about the operation of this universe than had ever been previously suspected. Because there remained unsolved problems, research went on, and this research—instead of producing simple solutions to the problems—led instead to the discovery of larger and more fundamental problems. And the ultimate solutions to these problems forced physical scientists to adopt a radical new view of the very nature of matter itself.

Many physicists of that era were unable to participate in the work we shall be describing, for they already knew a great deal about the universe, and tended to reject ideas that seemed to violate their good "common sense" too much. In this respect the reader of this book who is having his first encounter with physics has an advantage in approaching these concepts, for he does not have a great many long-accepted ideas about physics that have to be rooted out. Einstein, facing the objection that some of his theories violated good common sense, observed that, after all, common sense is only the "layer of prejudices we acquire before we are 16 years old."

The work of Albert Einstein has often been described as the start of twentieth-century physics, but it may equally well be considered

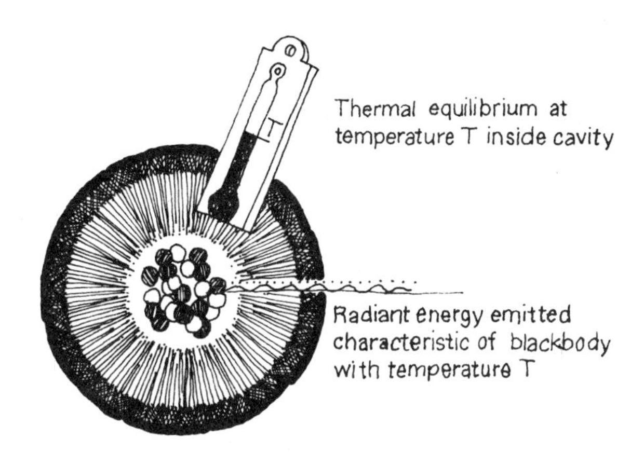

FIG. 6.1 An approximation of a blackbody.

as the capstone of classical physics. The ideas of relativity were foreign to the physics that had preceded them, yet the universe pictured by the classical physicist was one into which these ideas could be fitted and, once embedded, served to provide solutions to otherwise unsolvable problems.

After Einstein, classical physics had nowhere else to go. All the problems were not solved, and all the phenomena that had been observed were not explained, but the more researchers pushed forward, the more data they uncovered, the more the gap between experimental observation and theoretical explanation grew.

Of all those problems of nature that could not be explained in terms of classical physics, perhaps the most perplexing was the theory of radiation. Hot objects emit electromagnetic radiation. If they are hot enough, much of this radiation is in the visible part of the spectrum. Thus you may watch a bar of iron, as it is heated, become first a dull red, then glow a brighter red, and finally glow white-hot. Although we could study any heated object, it is convenient for us to consider radiation from an ideal radiator, otherwise known as a *blackbody* (Fig. 6.1).

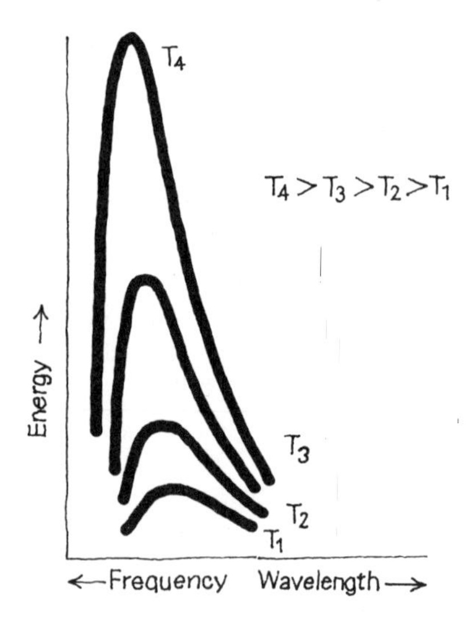

FIG. 6.2 Blackbody radiation.

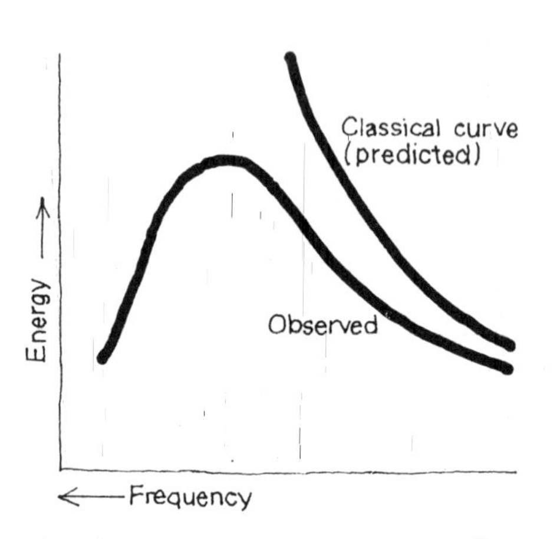

FIG. 6.3

The simplest arrangement that meets the criteria for an ideal radiator is a cavity, such as the inside of a sphere, which has only one small opening. The cavity and its walls, when heated to a given temperature, emit radiation from this opening. The observed distribution of the intensities of the various frequencies making up this radiation is represented in Fig. 6.2.

If the process of the emission of this radiation is understood, it should be possible to explain why the distribution curve has its characteristic shape. We are dealing with this radiation in terms of a frequency distribution, as if a number of individual oscillators inside the cavity were each radiating energy, each at a different frequency. The radiation spectrum that is observed is simply the sum of all this radiation emitted at various frequencies.

In the classical picture, there was invoked a principle known as the *equipartition of energy,* which stated that each individual oscillator has the same energy. Since the number of frequencies associated with the radiation process is without limit, and since each shares equally in the total energy radiated, it follows that the total amount of energy involved would necessarily have to be infinite. Thus the best that classical theory could do in explaining this radiation phenomenon was to predict that the energy distribution would appear as shown in Fig. 6.3. The high-frequency part of this curve increases without limit, a condition that was described as the "ultraviolet catastrophe." Obviously, classical physics totally failed to explain the process of the radiation of electromagnetic energy from a cavity.

In 1900 the German physicist Max Planck presented the solution to the problem of blackbody radiation. He suggested that instead of a continuous distribution of frequencies

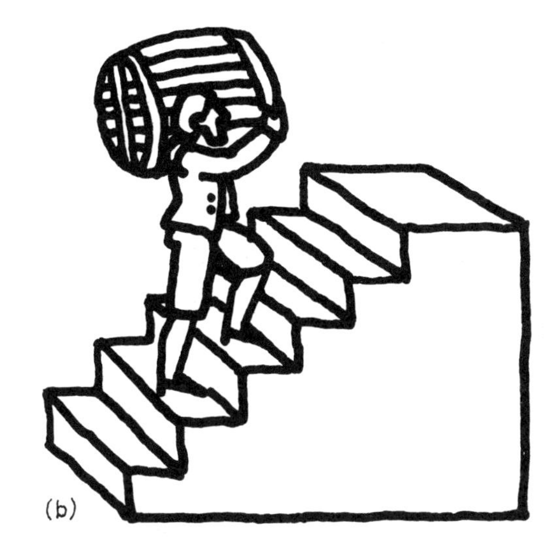

FIG. 6.4(a) Classical picture of available energies. Any energy level allowed. (b) Quantum picture of available energies. Only certain discrete energies allowed.

and energies, the oscillators responsible for the cavity radiation could emit (or absorb) energy only in discrete amounts, or *quanta* (from the Latin word *quantum*, meaning "how much"). Unlike the classical picture, in which each oscillator has the same energy regardless of its frequency, the energy content of each of Planck's quanta was directly proportional to its frequency. We now call the quantum of electromagnetic radiation the *photon*. The Planck relationship between energy and frequency is simply:

(Energy of a proton) = (a constant) × (photon frequency).

The constant used in the above relationship is now known as *Planck's constant*, and is written as h.

We might contrast the energies available in the classical picture and those given by Planck by thinking of the energy levels in the former case as being comparable to a smooth ramp, while the latter are like a set of stairs. In the classical picture any position on the ramp—any energy level—is available, while in the latter case one must step from one level to the next: only certain discrete energy levels are possible (Fig. 6.4). The emitted cavity radiation, as described by the Planck explanation, is the sum of the radiation emitted by all the oscillators in the cavity, each emitting radiation of a different frequency. Now the higher its frequency, the greater the energy possessed by any given oscillator. But Planck showed that the *number* of high-frequency oscillators is much smaller than the number of oscillators of lower frequency. Thus the extent to which the higher frequencies participate in the energy-sharing process is greatly restricted, and the ultraviolet catastrophe is avoided.

When calculations were made for cavity radiation using Planck's theory, the result was a

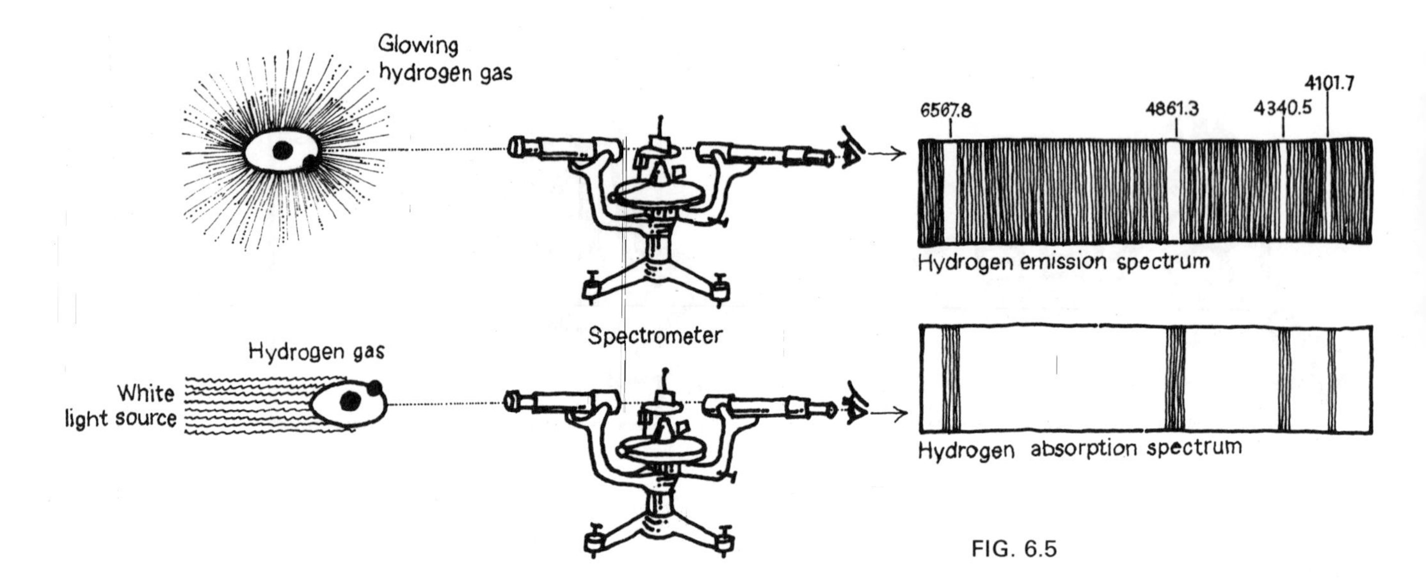

FIG. 6.5

theoretical energy-distribution curve for radiation that matched the experimental curve. Planck's idea had led to the discovery of the *quantization* of electromagnetic radiation.

6.2 QUANTIZED ENERGY LEVELS

The success of the quantum theory in explaining blackbody radiation led quickly to even greater triumphs for this idea that light is absorbed or radiated only in discrete energy packages. In 1905 Einstein was able to explain the emission of electrons from metals when light is incident upon them—known as the *photoelectric effect*—in terms of the absorption of incident photons by the electrons of the metal. But perhaps the greatest triumph of this idea was found in the work of the Danish physicist Niels Bohr, who first explained the role of electrons in the emission and absorption of light by atoms.

In Chapter 4 we learned that Rutherford's discovery of the atomic nucleus had led to a picture of the atom having electrons orbiting the nucleus. But, before the work of Bohr, the nature of these orbits was not really understood. Bohr's first work was with the simplest atom, hydrogen, with its lone electron.

It was known that the hydrogen atom, when supplied with energy, emits electromagnetic radiation. If hydrogen gas is confined to a glass tube and an electric discharge passed through it, the gas will glow (Fig. 6.5). To the eye this

Niels Bohr

light appears pink, although an examination of the emitted radiation with a spectrometer reveals that hydrogen emits only certain discrete frequencies of light. Conversely, if light having a broad spectrum of frequencies—white light—is passed through hydrogen gas that is not glowing, energy is absorbed by the gas at precisely the same frequencies as it is emitted by hydrogen gas that is glowing.

Bohr explained this phenomenon by saying that the electron of the hydrogen atom is not free to orbit the nucleus in any arbitrary orbit, but is restricted to certain allowed energy levels. This restriction is comparable to saying that an earth satellite may go into orbit at some distances above the earth but not at others. Each

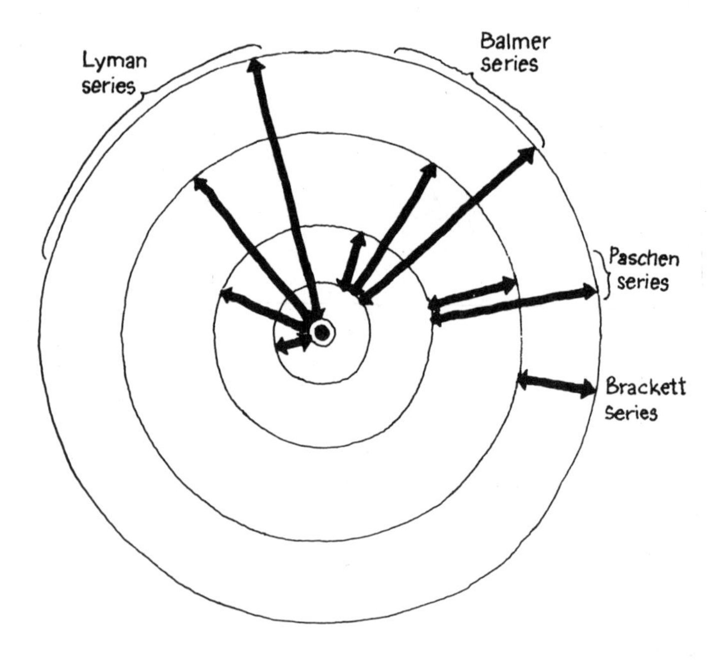

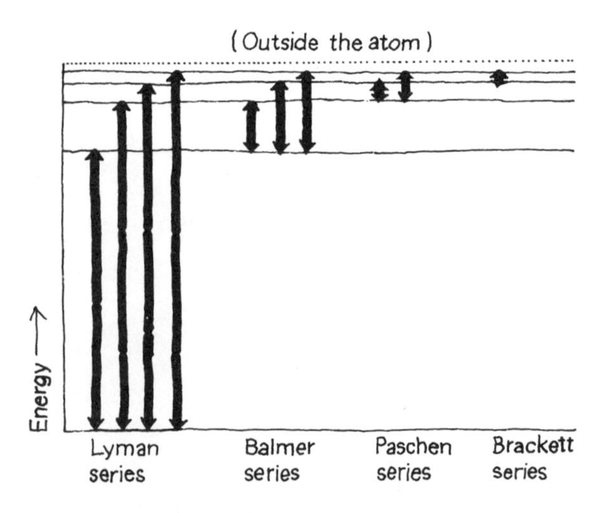

FIG. 6.6 The Bohr model of the hydrogen atom.

of the allowed electron orbits represents a different *energy level.* Thus an electron may jump from one orbit to another only by emitting or absorbing energy.

Bohr was able to explain the existence of the hydrogen spectral lines in terms of these discrete energy levels. Suppose that a hydrogen electron in its lowest energy level—known as the *ground state*—is excited to the next highest energy level in the atom. Electrons in atoms are like some people. They exist in the lowest energy state whenever possible, and our electron will soon return to the level from which it came. This electron may return to the ground state only by emitting a photon of energy equal to the difference in energy between the ground state and the excited state. Figure 6.6 shows the first few energy levels for this electron. The hydrogen lines representing electron transitions to a given energy level are identified as belonging to different *series* of lines. Those transitions involving the ground state are known as the Lyman lines, those to the first excited level are the Balmer lines, etc. The first four lines of the Balmer series are the only ones in the visible region of the spectrum. (The series of lines are given the names of early researchers in spectroscopy.)

Note that the spectral lines that may be emitted by an electron have frequencies precisely equal to the energy differences between the allowed levels in which this electron may be found. For the heavy atoms, in which many electrons are involved, the spectral patterns emitted or absorbed are highly complex, but each spectral line still represents the energy difference between two allowed electron energy levels. From an analysis of the spectral lines of a given element, it is possible to discover the allowed electron energy levels for that element. The spectrum of each element is unique, and spectral patterns provide a means of "fingerprinting" atoms.

6.3 WAVELIKE PARTICLES: THE deBROGLIE HYPOTHESIS

In 1925 a French graduate student, Louis deBroglie, presented a doctoral dissertation containing the extremely controversial hypothesis that every particle possesses wavelike properties. He proposed that the wavelength associated with a particle is equal to Planck's constant divided by the momentum of the particle. (The momentum of an object is equal to the product of its mass and its velocity.) Planck had said that light, which had been thought to be wavelike, has particle-like properties. Now deBroglie was suggesting that all particles have wavelike properties!

To see how the deBroglie hypothesis could explain the quantization of atomic electron energy levels, we shall first examine the vibrations of a string, such as a guitar string. When one picks a guitar string, it is displaced from its equilibrium position and set to vibrating in several modes. The simplest way the string can vibrate is with a single amplitude maximum at its center, while each end of the string remains, necessarily, stationary. The sound emitted by a string vibrating in this way is the lowest frequency the string can produce, and is known as its *fundamental* frequency. But the string may vibrate in other modes as well. Figure 6.7 illustrates the first few vibrational modes for a string. The frequencies emitted by the string vibrating in these modes are twice, three times, etc., the fundamental frequency,

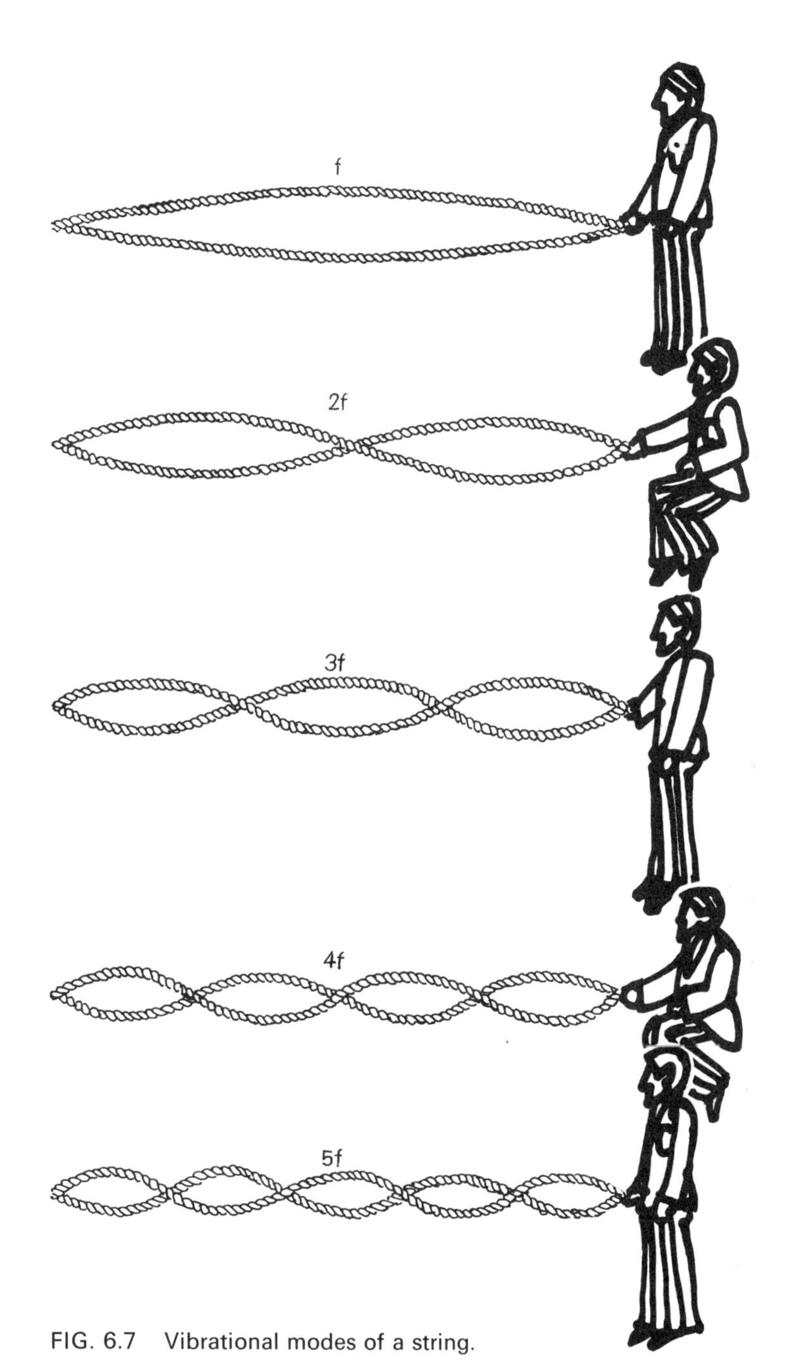

FIG. 6.7 Vibrational modes of a string.

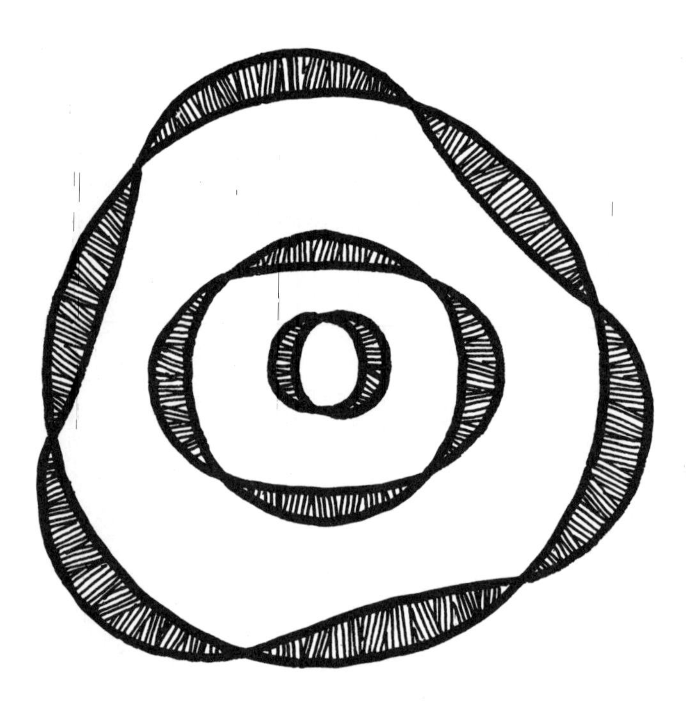

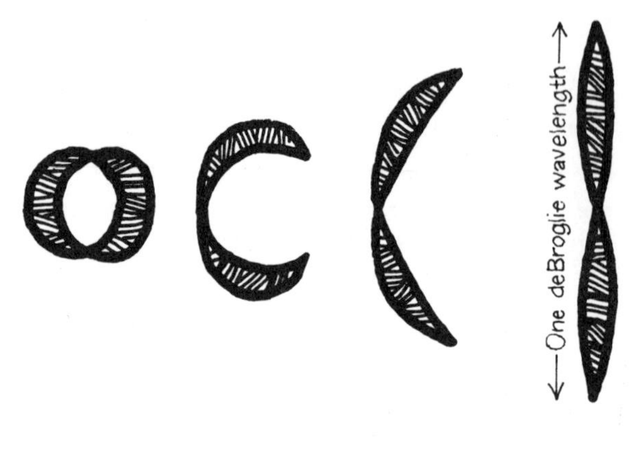

FIG. 6.8 Electron orbits as deBroglie waves.

and are known as *overtones* or *harmonics.* (The fundamental is also known as the first harmonic; thus the first overtone and the second harmonic are the same.) Since a real guitar string, when picked, vibrates in a mixture of the modes shown, the ratios of the amplitudes of the various harmonics are responsible for the quality of the sound produced. It is this characteristic variation in quality that enables one to distinguish between a middle A played on a guitar and the same note plucked on a violin.

An electron, as described by the deBroglie hypothesis, has a wavelength that depends on its energy. To illustrate how this is involved in the hydrogen atom, consider that the lowest energy level is made up of one electron wavelength wrapped precisely around the atom and joined onto itself (Fig. 6.8). Thus the ground-state energy level for the electron corresponds to the fundamental of the vibrating string. Exactly two electron wavelengths are joined together to form the next-highest energy level for the atom: the first excited state. Three wavelengths, fitted together, form the second excited state, and all higher energy levels are made up of more electron wavelengths. We may even use this picture to explain why the electron may not exist in any energy levels other than the ones that are allowed, for if the electron wave does not precisely join onto itself the

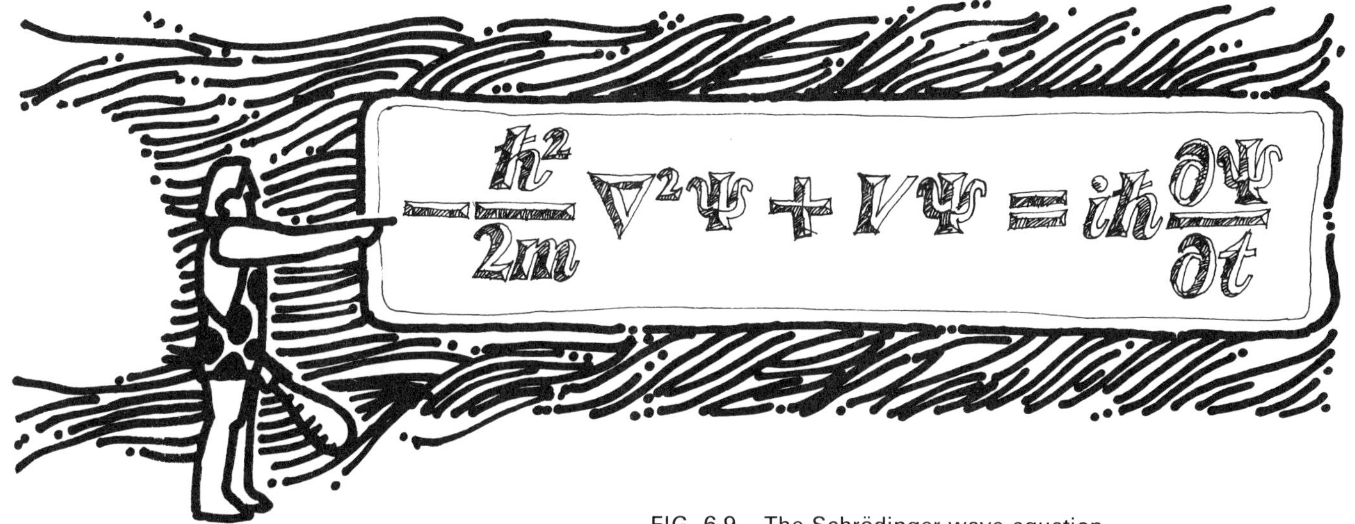

FIG. 6.9 The Schrödinger wave equation.

wave will be damped out—it will average to zero—after a number of orbits.

Thus the deBroglie hypothesis provided a satisfactory explanation for the quantization of the energy levels of the hydrogen atom. Of course deBroglie and his colleagues did not merely draw pictures and present hand-waving arguments or argue from analogy, but produced a precise mathematical statement of this theory. In 1926 an Austrian physicist, Erwin Schrödinger, formalized this theory into a partial differential equation, which has become famous as the *Schrödinger wave equation.* While we shall make no calculations using this equation, we present it for your inspection in Fig. 6.9. At about the same time Schrödinger's work appeared, the German physicist Heisenberg published a theory leading to identical results, but cast in a different mathematical form.

6.4 A PARTICLE IS A WAVE IS A PARTICLE

The deBroglie hypothesis applies to all particles, but—because of their convenient size and availability—electrons were most extensively used in investigating the validity of this theory. If electrons really do behave like waves in the hydrogen atom, should they not exhibit other wavelike properties? Specifically, waves may

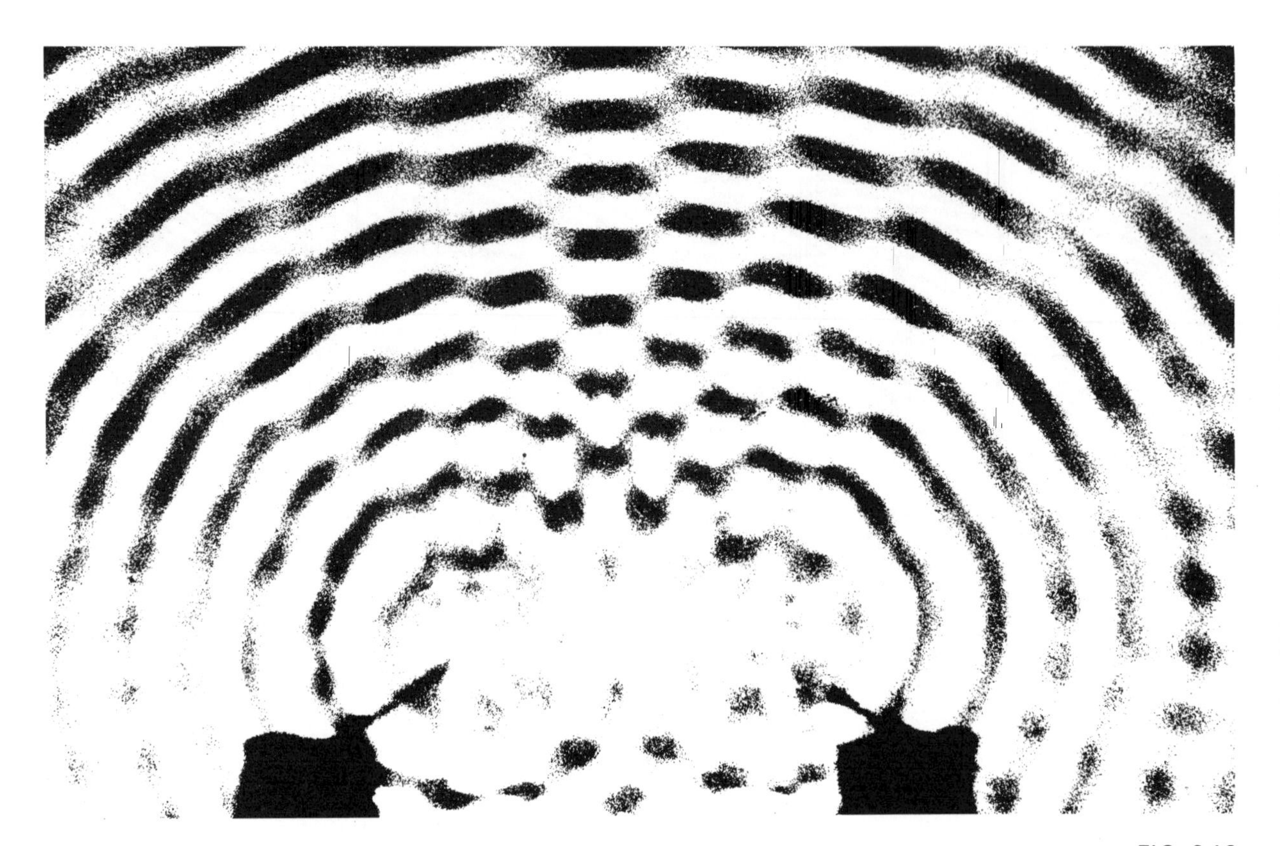

FIG. 6.10

interfere with one another, forming an interference pattern. Investigations started immediately to see whether electrons would produce the same effect.

You may see an example of an interference pattern if you drop two rocks into a calm body of water a few feet apart. Rocks of the same size dropped from the same height should form identical wave patterns in the water (Fig. 6.10). Where the wavefronts that go outward from the points at which the rocks were dropped into the water meet one another, interference patterns are observed. At points where two wave crests come together, the interference is constructive: the two crests add to produce a wave that has twice the amplitude of either wave. Where the crest of one wave meets the trough of another wave, the interference is destructive: the waves cancel each other out at that point. Thus a series of wavefront maxima and minima are produced by the interference of the two wavefronts.

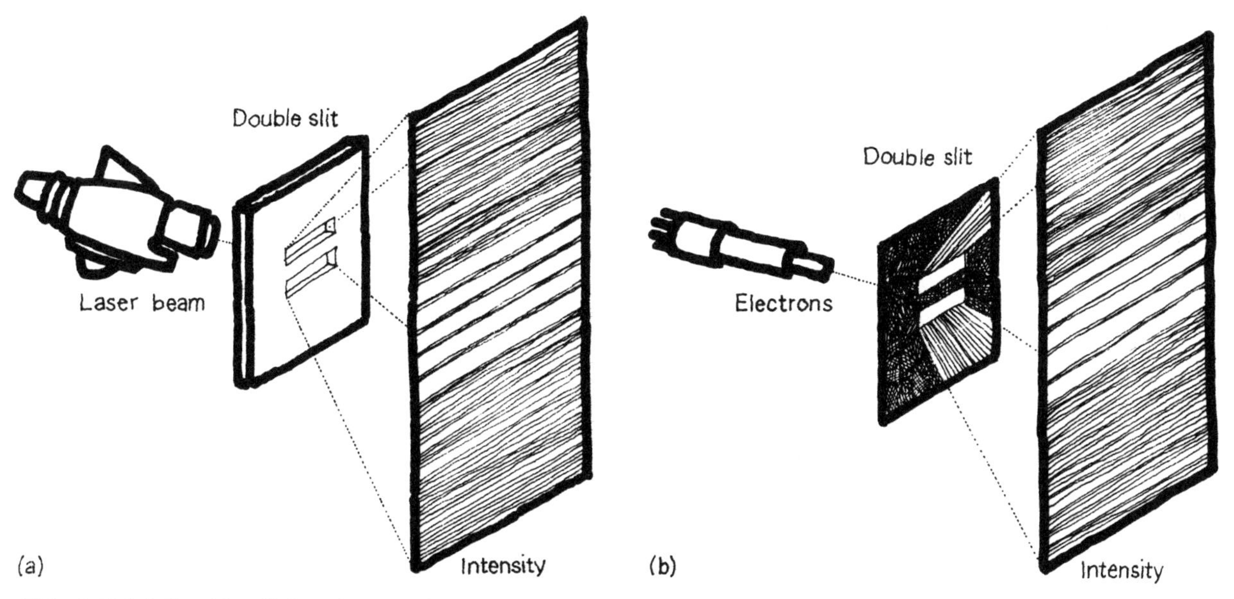

FIG. 6.11 (a) Double-slit interference (light). (b) Double-slit interference (electrons).

The same phenomenon may be observed when light passes through a pair of narrow openings. An experimental setup that produces this effect is illustrated in Fig. 6.11. Light, from a laser, may reach a screen by taking either of two paths—via one of two slits—through a mask. The intensity of the light reaching the screen will be due to the light waves that arrive by each of the two possible available paths. At the very center of the screen, the two light paths are identical. Therefore the same number of wavelengths of light is involved whichever route the light takes. The two beams arrive precisely in phase, there is constructive interference, and the intensities add together to produce a bright spot on the screen.

A short distance away from the central bright area of this pattern, there are dark areas. A dark area is located at the place at which the two paths of the arriving light differ by exactly half a wavelength. Here the two beams interfere destructively and cancel each other out. This argument may be repeated for locations up and down the screen, and, it should be obvious, the screen will contain alternating bright and dark areas. This striped pattern is known as an *interference pattern,* and it simply indicates that the light waves that arrive at the screen by two different paths may interfere with each other. The fact that two paths are involved is crucial in producing this pattern, for if one slit were to be covered, the interference pattern that had been seen would disappear.

If light waves may exhibit an interference pattern and electrons have wavelike properties, should not one be able to observe the same pattern with electrons? The answer to this question provided one of the crucial tests of the de-Broglie hypothesis. Because of the difficulty in performing the experiment, a test using electrons and a double slit was not actually done until 1961. But in 1927 two American physicists, Davisson and Germer, provided an experimental verification of the dual nature of electrons by observing diffraction* effects when electrons were scattered from the surface of a metal crystal. Their experiment demonstrated that electrons may indeed act like particles or waves.

The implications of this wave/particle duality for the electron are another of the violations of "common sense." When one imagines a beam of electrons incident on a double-slit arrangement, it surely appears logical that while some electrons may go through one opening and some through the other, each electron is thought of as going through one slit *or* the other. Yet when the experiment is performed, it is observed that only when both slits are open and available to each electron will the interference pattern be seen. The intensity of the electron beam at the slits may be reduced until it is certain that only one electron at a time arrives at the slits, yet the interference pattern will continue to be seen so long as both slits remain open. If one slit is closed, the alternating maxima and minima in the observed intensity pattern vanish.

What happens is not that one electron interferes with another electron; rather this effect involves each electron by itself. The electron can no longer be thought of as a small billiard-ball-like sphere, but as being represented by a *wave packet* which has some finite extension in space. When both slits are available to this wave packet, interference effects follow, just as they did when electromagnetic waves were incident on a double slit. If one slit is closed, the electron wave behaves like a light wave in a similar circumstance.

Could the same interference patterns be observed for larger objects? For example, think of firing machine-gun bullets against a steel plate with two narrow openings, or throwing baseballs against a similar barrier. Interference phenomena are not observed with bullets or baseballs, simply because their wavelengths are so short that the effects are too small to be observed. Remember that the deBroglie wavelength of any object is equal to Planck's constant divided by the product of the mass and velocity of the object. Planck's constant is so small a number that the wavelength of an object weighing 1 gram and traveling at 1 centimeter per second is only about 6×10^{-27} centimeters.

*Diffraction is a complicated case of interference.

6.5 QUANTUM-MECHANICAL MEASUREMENTS

Now that we have started to think of very small objects, such as electrons, as exhibiting wavelike as well as particle-like properties, we shall find that many of our classical ideas about making measurements must be revised. So long as one thinks of an electron as a tiny, hard sphere, one naturally thinks of describing its location precisely. A classical electron could be located at such-and-such a point in a convenient coordinate system, for example. But an electron that is described by a wave packet cannot be located with such certainty.

In making quantum-mechanical measurements, one is able to express only the *probability* of finding an object at the point described. For example, if a classical electron were placed in a closed box, at some later time you would be able to say with 100% certainty that the electron is still inside the box. This is no longer true in the world of quantum mechanics, for unless the barriers of the box are infinitely high, there is a finite probability—however small—of the electron being found *out*side the box. This follows from the fact that the electron wave packet will not all be confined to the box, but will have some finite amplitude outside the box. So long as a particle wave function is nonzero at a point, there is a finite probability of locating the particle at that point.

Another difference between quantum-mechanical and classical measurements is discovered when one attempts to measure quantities simultaneously. In classical physics, aside from the experimental problems involved, there is nothing conceptually impossible about simultaneously measuring both where a particle

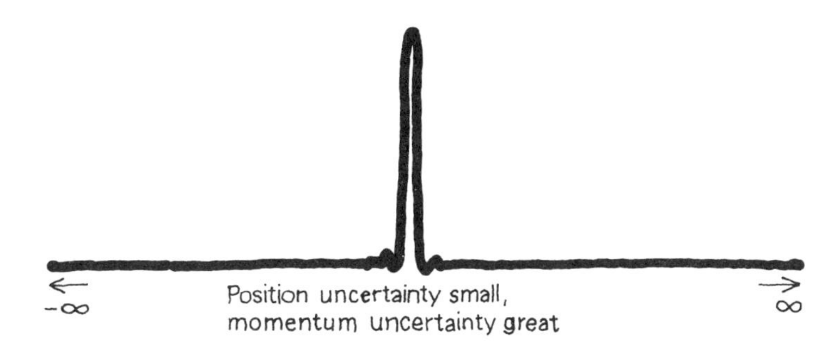

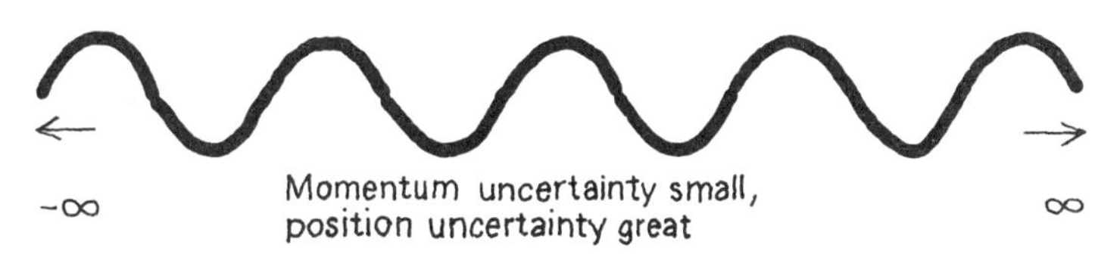

FIG. 6.12

is located and how fast it is traveling. But in quantum theory, the precision of a simultaneous measurement of both the position and the momentum of a particle is limited, so that the product of the uncertainty in making the position measurement and the uncertainty in making the momentum measurement is on the order of magnitude of Planck's constant (Fig. 6.12). This result, stated formally, is the *Heisenberg uncertainty principle.*

We may illustrate why the uncertainty principle holds by noting that if the position of an electron is to be weighed very precisely, its wave packet must be reduced until it occupies only a very small volume in space.

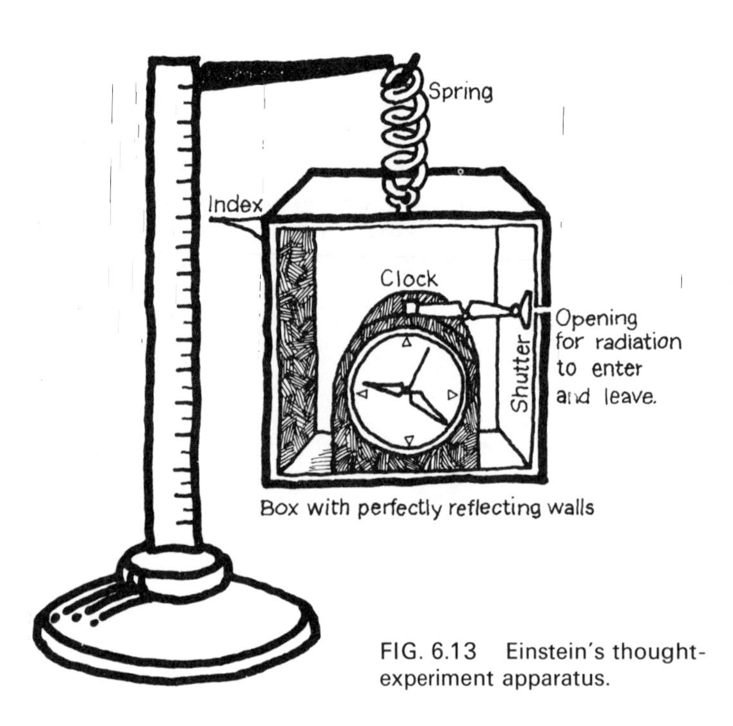

FIG. 6.13 Einstein's thought-experiment apparatus.

To see how this may be done, we may think of the electron wave packet as being made up of a great many pure sine waves of different frequencies and amplitudes that are added together to make up the electron wave packet. These sine waves may be added in such a way that they cancel each other out throughout all space, except in the very small volume in which the electron is located. The electron thus described may be very precisely located, but we shall now be placed in the position of knowing virtually nothing about the *momentum* of that electron, for the momentum of an electron may be specified precisely only when its wave packet contains a single pure sine wave.

Since our located electron wave packet is made up of a great many sine waves of different frequencies, we know almost nothing about its momentum. Conversely, an electron for which we knew the momentum very precisely would be represented by a single pure sine wave of constant amplitude; but there would be equal probabilities of finding the electron anywhere from here to infinity.

The Heisenberg uncertainty principle tells us that the product of the uncertainty of a particle's position and the uncertainty of its momentum in that direction is of the order of Planck's constant for each of the three spatial dimensions. Since time has a status comparable to that of the spatial dimensions, the uncertainty relationship in this fourth dimension is stated: The product of the uncertainty in the measurement of energy and the uncertainty in the measurement of time is of the order of magnitude of Planck's constant.

If you have found that grasping some of these quantum-mechanical ideas seems to do violence to your "common sense," you might be made somewhat happier to know that even the great Einstein was never too comfortable with some of the ideas of quantum mechanics. In 1930 a number of physicists were in Brussels to attend an international conference on the problems of the then-still-new quantum theory. At that conference Einstein proposed a "thought experiment" which he felt would contradict the time-and-energy uncertainty relationship (Fig. 6.13).

Einstein's experiment involved the construction of an imaginary box, its interior walls being lined with perfectly reflecting mirrors, and having one small opening. This aperture would be opened or closed by the action of a shutter connected to a clock inside the box. Einstein proposed that the shutter be opened, the box filled with radiation, then the shutter be closed again. The box would now be weighed. At a preselected time, the clock would open the shutter, an amount of radiant energy would escape from the box, then the shutter would be closed again. The box would be weighed a second time, and, from the difference in weights, the exact amount of energy that escaped would be calculated. (This calculation would be made using the Einstein relation, $E = mc^2$, for the box would be lighter after some energy had escaped from it.) Because the weighings could be performed to any desired tolerance, and since the clock could operate to any desired degree of accuracy, the uncertainty condition involving energy and time would be violated, because the uncertainties in both energy and time could each—in principle—be made equal to zero.

One of those in attendance at the conference was Niels Bohr. Apparently he got little sleep that night, trying to solve the Einstein puzzle, for it has been reported that he appeared a bit bedraggled the next morning—but he had the solution to the problem. He pointed out that, if the energy that escaped the box were to be *weighed*, the box would have to move vertically in the earth's gravitational field. From Einstein's own General Theory of Relativity, it can be shown that a clock runs more slowly in a large gravitational field than in a smaller one; thus the movement of the box in the earth's gravitational field would introduce a small uncertainty in the timing of the shutter. Further, since the box would be set in motion during this operation, the uncertainty relations involving position and momentum could be applied. Bohr used these equations, along with one of Einstein's greatest discoveries, to defeat Einstein's own argument.

6.6 THE EXCLUSION PRINCIPLE

By now it should be rather apparent that when we begin looking into events that happen to things of smaller and smaller sizes, we do not just find "more of the same" of what we see on the scale to which we are accustomed. Instead, as we have explored the microuniverse in which quantum mechanics reigns, we have discovered things like the uncertainty principle which may offend our ideas of "common sense." We want to look now at one more quantum-mechanical phenomenon, which is of the utmost importance in explaining the strikingly different chemical and physical properties of the elements.

So far in this chapter we have seen two different pictures of the atom, but both have illustrated hydrogen, the simplest of the elements. Hydrogen's one electron normally occupies the lowest energy level in the atom, which is the allowed orbit nearest the nucleus. But where are the electrons located in the more complex atoms?

A naive assumption would be that the electrons of the heavier elements are also normally found in the first allowed orbit. This would mean that uranium would have 92 electrons orbiting in the "airspace" where hydrogen has just one! Actually, the electrons of the more complex atoms are required to occupy *higher* energy levels, because of a principle, first described by the Austrian physicist Wolfgang Pauli, that is usually known as the *Pauli exclusion principle.*

The Pauli principle may be used to tell where

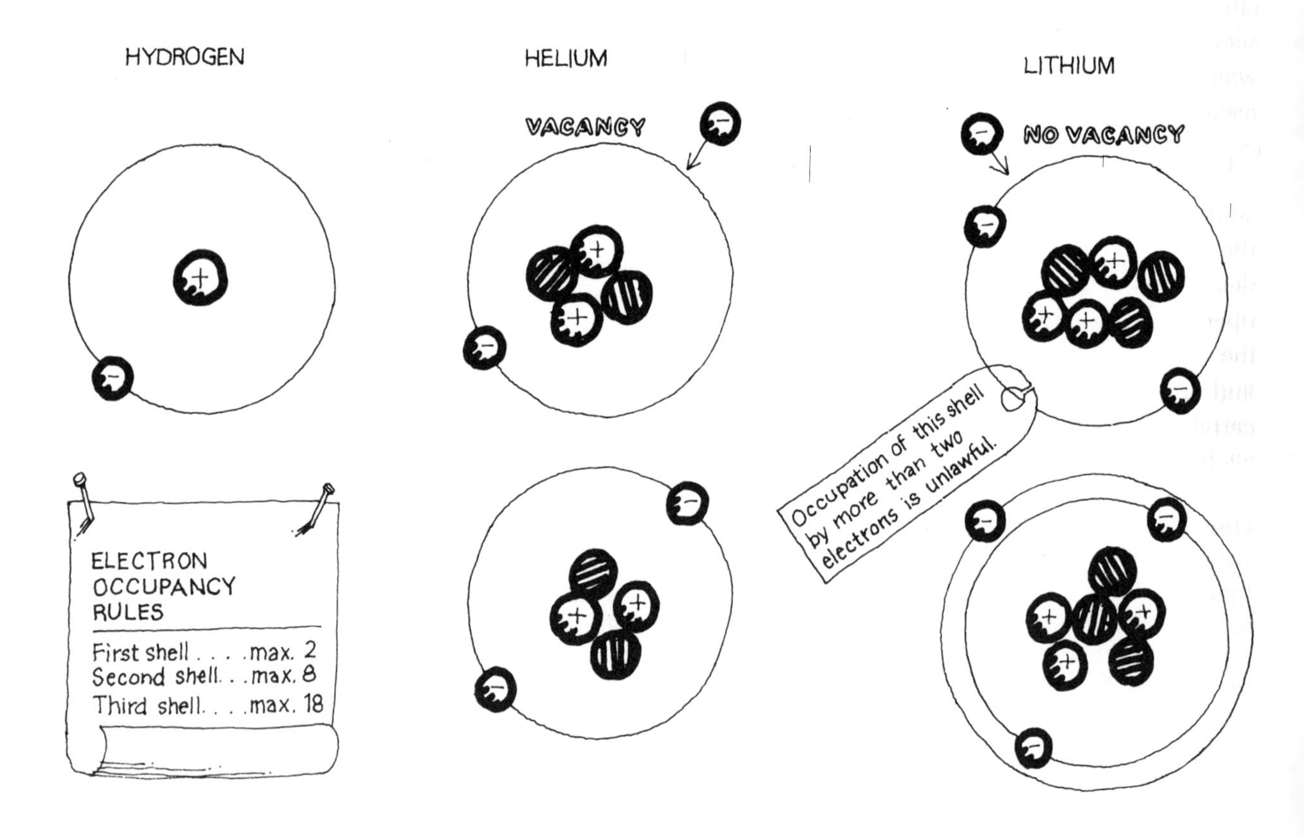
HYDROGEN
HELIUM
LITHIUM
VACANCY
NO VACANCY
ELECTRON
OCCUPANCY
RULES
First shell . . . .max. 2
Second shell. . .max. 8
Third shell. . . .max. 18
Occupation of this shell by more than two electrons is unlawful.

FIG. 6.14

electrons *can't* stay in an atom. Electrons are identical particles. You can't hang a tag on an electron and identify it later, since any two electrons may swap places without your being able to tell the difference. But, although electrons are identical, the *states* they occupy in the atom are not. It happens that we may describe an electron state in an atom by a combination of four numbers, which are known as the *quantum numbers,* representing a particular electron state. The Pauli principle, then, simply states that in a given atom *no two electrons may have the same set of quantum numbers.*

Each set of quantum numbers is associated with an electron orbit. It turns out that there are only 2 sets of quantum numbers that may be assigned to the smallest electron orbit, 8 to the next orbit, 18 to the next, and so on. These electron orbits are known as *shells,* and each of the more complex atoms is built by sequentially assigning electrons to the lowest energy orbit that is vacant (Fig. 6.14).

Helium has only two electrons, so both are able to stay in the first shell. Lithium has three electrons; consequently two are in the first shell, while the third must occupy the second shell. Beryllium has two electrons in the first shell and two in the second.

Each shell also contains subshells, which may be filled even though the whole shell is not. Electrons in filled shells or subshells do not participate in chemical reactions as do electrons in unfilled shells. The inert gases—helium, neon, argon, krypton, and radon—have electrons only in closed shells or subshells. Elements with the same numbers of electrons in unfilled shells, such as carbon and silicon or calcium and strontium, are chemically similar.

6.7 THE QUANTUM REVOLUTION

The concepts of the quantum theory have, in a very real sense, revolutionized our view of the physical world. Just as the theories of relativity profoundly changed our ideas about space and time and mass and energy, quantum mechanics has uprooted many old ideas of waves and particles and what is subject to measurement on very small scales.

The quantum theory was first needed because of the utter failure of classical theories to explain certain phenomena. Once the secrets of the quantum world began to be unlocked, many new phenomena—also totally inexplicable in classical terms—were found.

Yet it must be understood that, although quantum mechanics has provided explanations of nature where Newtonian mechanics completely failed, it is incorrect to think of Newtonian mechanics as having been "replaced" by quantum theories. Instead, classical mechanics has simply been recognized as a limited application of a more complete physical theory.

We found that the equations we have always used for lengths and times are perfectly satisfactory so long as we are not involved with objects traveling near the speed of light. Similarly, classical equations will continue to be used to make calculations in those areas in which they provide a satisfactory picture of what is happening. If we wish to calculate the orbit of a satellite, for example, we still use classical-mechanical calculations.

However, it is equally wrong to say that no quantum-mechanical effects are observable on the scale in which we live. Just as $E = mc^2$ describes a relationship between energy and mass that is important whether one is traveling near the speed of light or not, applications of quantum-mechanical phenomena are seen daily in laser operation and solid-state electronics. The world in which we live *is* a world of quantum mechanics.

FOR MORE INFORMATION

Adolph Baker, *Modern Physics and Antiphysics,* Reading, Mass.: Addison-Wesley, 1970, pages 123–229. The sections "A wave is a particle" and "A particle is a wave" may be particularly helpful.

Victor Guillemin, *The Story of Quantum Mechanics,* New York: Scribner's, 1968. A most illuminating treatment of quantum mechanics for the general reader. Both history and philosophical implications are discussed.

George Gamow, *Thirty Years that Shook Physics: the Story of Quantum Theory,* Garden City, N.Y.: Doubleday, 1966. An account of the development of quantum mechanics, including many anecdotes about the persons involved, by one who was there when it happened.

Louis deBroglie, *Matter and Light: the New Physics* (translated by W. H. Johnston), New York: Dover, 1939. A collection of articles and lectures, including deBroglie's address when he received the Nobel prize. (In paperback.)

J. Andrade e Silva and G. Lochak, *Quanta* (translated by Patrick Moore), New York: McGraw-Hill, 1969. (In paperback.)

QUESTIONS

1. What was the apparent status of our understanding of the laws of physics at the close of the nineteenth century?
2. Describe two observable changes in the radiation from a blackbody as that body gets hotter.
3. What is the principle of equipartition of energy?
4. How does the observed blackbody radiation curve differ from the classical prediction of how it should look?
5. What was the ultraviolet catastrophe?
6. What was Planck's solution for the ultraviolet catastrophe?
7. What does the symbol h represent?
8. How do the emission and absorption spectra of a single element differ?
9. Describe Bohr's picture of the atom.
10. What do Lyman, Balmer, Brackett, and Paschen have in common?
11. What was the deBroglie hypothesis?
12. How did deBroglie's hypothesis fit into Bohr's picture of the atom?
13. What did Schrödinger contribute?
14. What is interference in waves?
15. What was the contribution of Davisson and Germer?
16. What does a wave packet have to do with an electron?
17. State the Heisenberg uncertainty principle.
18. Name some pairs of quantities that can never be simultaneously measured to any desired degree of accuracy.
19. Why aren't the quantum-mechanical properties of objects the size of baseballs apparent?
20. Describe the Einstein-versus-Bohr thought experiment.
21. Compare the quantum numbers of electrons to Social Security numbers.
22. Name some chemically similar elements. Why are they similar?
23. Has quantum mechanics "replaced" classical mechanics?

PROBLEMS

1. If Planck's constant were only large enough, quantum-mechanical effects would be observable for baseball-sized objects. Describe how a baseball game might be played in a universe with a large value of h. (If you aren't familiar with baseball, use, in your answer, any game played with a ball.)

2. There is a theory that our solar system is only an atom in a super-universe, with the planets playing roles of electrons and our sun as nucleus. Support or attack this theory.

3. Einstein has been quoted as having said, in connection with the uncertainty principle, "God is subtle, but not cruel." What do you think he meant?

4. A Newtonian universe has sometimes been described as being completely deterministic: if you knew the description, position, and momentum of every particle and the forces acting on it at any given time, you could accurately predict what would happen at any other given time. If this is so, what fundamental difference exists between a quantum-mechanical universe and a Newtonian universe?

5. What difficulties would one encounter in precisely predicting events in a Newtonian universe? (See Problem 4.)

6. Compare the status of physical knowledge of the universe today and attitudes concerning it with the status and attitudes of the last part of the nineteenth century. How do you suppose the physical scientist of the twenty-first century will view your answer to this question?

7. It has been observed that, more often than not, a physicist's most important contributions have been made shortly after he has completed his formal training and begun to work in a field of physics. Can you suggest why this has been particularly true in quantum mechanics?

8. Descriptions of quantum-mechanical phenomena that occur on a scale so small that they are difficult or impossible to observe directly are not likely to be very convincing to an unbeliever. How could you convince a skeptic that quantum effects really occur?

9. There is a view of nature which sees an "inward infinity" as well as an outward infinity. In this theory it is supposed that physical phenomena are similar whatever the scale. Hence the verse:

Larger fleas have smaller fleas,
Upon their backs to bite 'em;
And these in turn have lesser fleas,
And so on, *ad infinitum!*

Is this true? Why, or why not?

CHAPTER 7 THE BUILDING BLOCKS OF THE UNIVERSE

7.1 ELEMENTARY PARTICLES

The discovery of the neutron, in 1932, increased the total number of known particles to three. The electron had been identified in 1897 by Sir J. J. Thomson and the proton had been recognized as a constituent part of the Rutherford nucleus. Although no one at that time was quite ready to predict that not more than three elementary particles existed, surely it was never dreamed that one day the number of known "particles" would stand at approximately 200!

The first addition to the list of particles came about in the process of solving the troublesome problem of how the atomic nucleus is held together. One should expect, in a nucleus containing positively charged protons and uncharged neutrons, that the protons would repel each other and cause the nucleus to be highly unstable. Yet nuclei exist and some are extremely stable. What is the "nuclear glue" that binds the like-charged protons together?

Hideki Yukawa, a Japanese physicist, sought to answer this question in 1935 with his proposal that the force that binds the nucleus together involved a new and undiscovered particle. The force Yukawa described holds the nucleus together through an attractive *exchange force* that is *charge independent.* By charge independence we mean that this force binds together two protons, or two neutrons, or a proton and a neutron, in the same way. Because the electric charge of a particle makes no difference to this force, we may describe both protons and neutrons simply as *nucleons.*

It is easier to illustrate a repulsive exchange force than an attractive one. Suppose that there were two men, each wearing ice skates, facing each other from a few feet apart on a frozen pond (Fig. 7.1). Let one man weigh 180 pounds and the other 150 pounds, but let the 150-pounder hold a 30-pound medicine ball. Starting with both men standing still, let them throw the ball back and forth between them. When the 150-pound man throws the ball for the first time he finds that he is propelled backward when he throws the ball forward, since his total momentum was zero before he threw the ball and it must be zero afterward. The man who catches the ball is accelerated backward as he catches the ball, for his momentum, too, must be conserved; and he is accelerated backward again when he throws the ball. As the men continue to catch

FIG. 7.1 Repulsive exchange force.

and throw the ball, each is accelerated backward each time he catches the ball and again when he throws it. After the ball has been tossed back and forth several times, let the 150-pound man catch it and keep it.

At the start of the above process we had one 180-pound man and one 150-pound man who carried a 30-pound medicine ball, each stationary on the ice. At the end of the exercises, a 180-pound man and a 150-pound man who carries a 30-pound object are moving rapidly across the ice away from each other.

The process we have just described, if viewed by an observer far enough away that he could not see the medicine ball, would appear to be one in which the two men are *repelled* by each other and are each accelerated backward away from one another. In this example the medicine ball serves as the unseen exchange particle.

The theory offered by Yukawa involved an attractive, rather than a repulsive, exchange force that binds nucleons together. The exchange particle he proposed should have a mass about 200 times that of the electron and would be very short-ranged. Two years after Yukawa's suggestion, a particle of about the right mass was found, but its interaction with nucleons was found to be only one millionth of a millionth (10^{-12}) as great as that which the predicted particle should have with nucleons. The particle that had been found instead of the Yukawa particle was the μ-meson, or *muon*. (The muon is the particle we discussed in Chapter 3, in connection with relativistic time dilation.) The Yukawa particle, the π-meson, or *pion*, was not actually discovered until 1947.

The pion interacts with nucleons via a type of interaction that is different from any of the others we have yet described. Four types of interaction have been observed in our universe: the *gravitational, electromagnetic, "weak,"* and the *strong*—or nuclear—interactions. The gravitational interaction is the weakest of these four forces of nature. The electromagnetic interaction is second in strength to the nuclear force, while the strongest of these interactions is the nuclear force.

The weak force—the one associated with beta decay—is actually stronger than the gravitational interaction, though weaker than the electromagnetic force. As early as the 1930's it had been noticed that there appeared to be a bit of energy missing in the beta-decay reactions. Wolfgang Pauli postulated that this missing energy was carried away by a particle having no electric charge and interacting with matter only very weakly. Pauli had suggested that this particle be known as a "neutron," but after the discovery of the particle now known as the neutron by Chadwick in 1932, it was realized that the Pauli "neutron" was much smaller than its namesake. Asked about the differences in the two particles, Enrico Fermi described the Pauli particle as a *neutrino*, using the diminutive form in Italian (which Fermi was speaking at that time). The name stuck, and the neutrino—which interacts with matter so slightly that on the average one neutrino can pass though a plate of iron that is 100 light years across without interacting with anything—was finally discovered in 1955.

After high-energy particle accelerators became available, the number of known particles increased rapidly. These "atom smashers" produced particles in such profusion that the number of "particles" now known stands at approximately 200. Table 7.1 lists some of the better-known particles and their prominent characteristics.

The elementary particles may be classified in several ways. We may first divide them into the *hadrons*, which interact by the strong interaction, the *leptons*, which interact by the weak interaction, and the *photon*, the electromagnetic-interaction particle, which stands all alone in its family.

The hadrons are further divided into *baryons* and *mesons.* The number of baryons must remain constant in any sort of interaction: If there were 2 baryons involved in a reaction before it started, there must be 2 baryons left afterward, though they need not be the same baryons. The baryons are further divided into *hyperons* and *nucleons.* All the hyperons have masses greater than the nucleons. Hyperons are known as *strange* particles, while the nucleons are not strange. (You see, particle physicists speak a most exotic tongue!)

The mesons include the K-meson (or kaon), which is strange, and the pions, which are not.

The lepton family includes the muon, the electron, and two different kinds of neutrino, one associated with the muon and one associated with the electron. Among the great mysteries of particle physics is why the muon should be so very much like the electron, only 207 times as massive. However, in any interaction involving neutrinos, the muon neutrino and the electron neutrino act as quite different particles. The number of leptons is also always conserved.

7.2 ANTIMATTER

Symmetry is a very important thing to physicists. Many of them were rather unhappy to learn that the proton's mass is 1836 times as great as the electron's mass. Should not the basic positive and negative charge carriers have equal masses? Symmetry considerations, in fact, led P. A. M. Dirac, in 1928, to predict the existence of a particle having the same mass as the electron, but carrying a positive charge. This particle, the *positron,* was discovered in 1933—the first known antiparticle.

Note that there is a "mirror image" shown for all but two of the particles listed in Table 7.1. These images represent *antiparticles.* A particle and its antiparticle annihilate one another when they come in contact. All that remains of their encounter is the energy possessed by the particles, and this is radiated away. The inverse of this process also occurs: Energy may be converted into matter with the creation of a particle and its antiparticle.

One of the most common examples of the energy-to-matter conversion processes involves the electron and its antiparticle, the

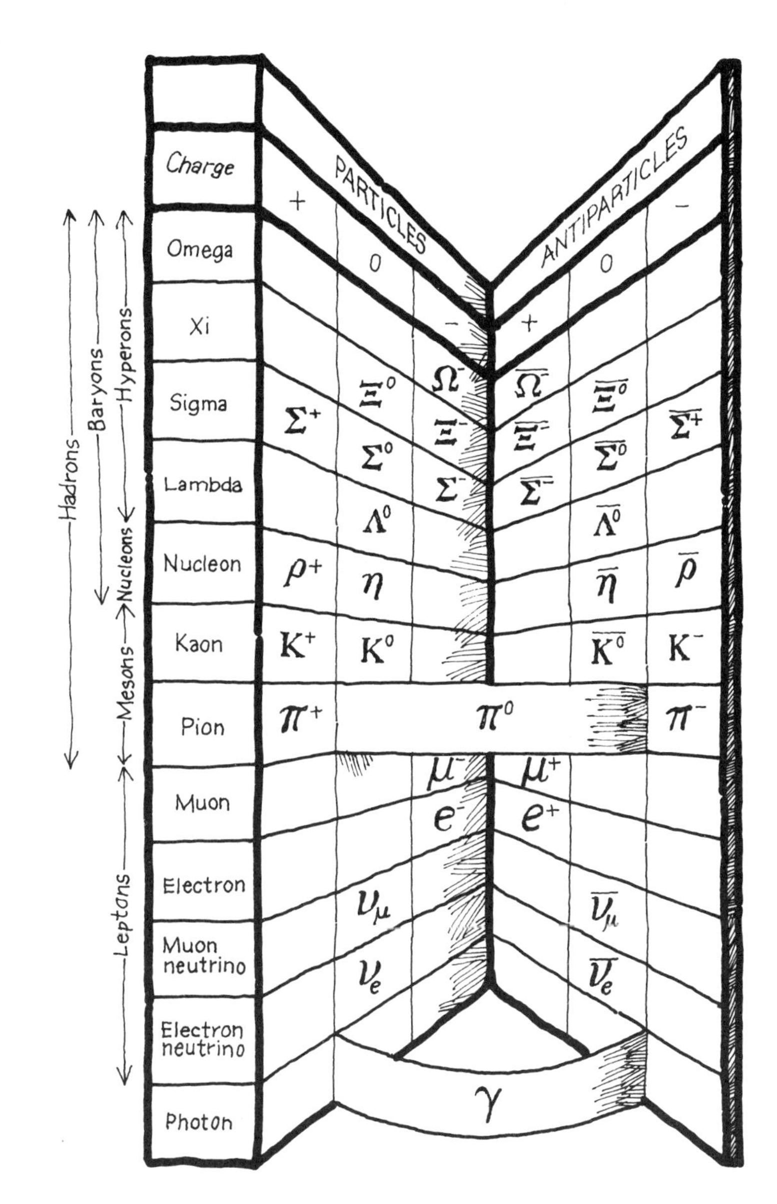

TABLE 7.1

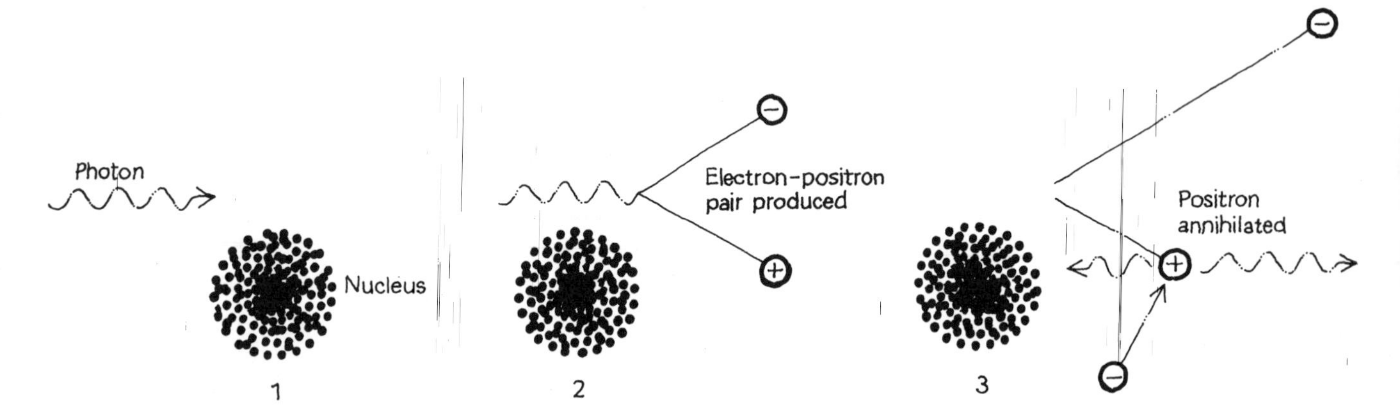

FIG. 7.2 Pair production.

positron (Fig. 7.2). When an electromagnetic quantum of sufficient energy is near an atomic nucleus, it may be converted into an electron and a positron in a process known as *pair production*. The positron created in this encounter doesn't last long, because there are lots of electrons around and, as soon as the positron encounters one, the particle-antiparticle pair annihilate each other, leaving behind an amount of energy equal to that which the particles had at the time of the encounter.

Of course creation and annihilation interactions are not limited to electrons and positrons; others involve protons and antiprotons, neutrons and antineutrons, etc. It is easy to think of an entire antiworld made up of antielements such as antihydrogen, which consists of an antiproton orbited by a positron. One cosmological theory suggests that there are equal amounts of matter and antimatter in our universe. The encounter of a large-scale matter object with one made of antimatter would lead to an annihilation and energy-production process far more efficient and energetic than the hydrogen bomb.

In every case in which a particle carries an electric charge, its antiparticle carries the electric charge of opposite sign. When one is making calculations involving the conservation of numbers of baryons or leptons, one counts an antiparticle in the balancing equation as minus one particle. Two particles—the uncharged pion (referred to as a "pi nought") and the photon—are their own antiparticles.

7.3 CONSERVATION RELATIONSHIPS

Those quantities that remain constant during physical interactions are said to be *conserved*. The fact that they do remain constant enables us to use them in making useful calculations involving these interactions. In the early stages of your encounter with physics, you might have wondered about some of the quantities physicists deal with. For example, what's so magical about linear momentum, the product of a particle's mass and velocity? The answer is that the total linear momentum of any system is always conserved so long as no external forces act on the system. This fact is especially useful when one is making calculations for rockets and space-vehicle propulsion systems.

A number of other quantities also appear to always be conserved. Among these are the total *mass-energy* of a system, its *angular momentum* (in the absence of external torques), and its total *electric charge*. We noted in the previous section that the number of baryons and the number of leptons in a system are also conserved. Strangeness is conserved for strong and electromagnetic interactions, though not in reactions involving the weak force.

Before 1956 it was thought that *parity* is always conserved. The conservation of parity means that the mirror image of any real physical experiment is also a valid physical experiment. In that year several experimenters found evidence that parity is not conserved in weak interactions. From the time of the discovery of nonconservation of parity until 1965, it was believed that the mirror image of any experiment is still valid if all the particles involved are exchanged for their antiparticles. The exchange of a particle and its antiparticle is known as *charge conjugation*, and this combination of parity and charge-conjugation invariance is referred to as *PC symmetry.*

In 1965 a group of physicists working at the Brookhaven National Laboratory found evidence for violation of PC symmetry. It is still believed that if one also performs a time reversal (the velocities and rotations of all particles are reversed) on an interaction for which there is PC violation, the result will again become a possible physical experiment. This overall symmetry, abbreviated as *PCT invariance*, is thought to hold for all interactions.

7.4 P, C, AND T

The belief that PCT invariance is valid for all interactions comes from a fundamental theorem in quantum field theory. If the PCT theorem is found to be invalid, some changes in the basic concepts of theoretical physics must come about.

Parity conservation implies that if a real physical event is viewed in a mirror, the mirror image will also represent a possible physical event. Now a mirror image is different from the object reflected in some important respects: if you are right-handed, your mirror image is left-handed. Similarly the mirror image of a right-handed screw is a left-handed screw.

The experiment that first demonstrated the nonconservation of parity was proposed in 1956 by T. D. Lee and C. N. Yang and performed by Mrs. C. S. Wu. The experiment suggested by Lee and Yang involved an observation of beta decay in cobalt 60. When the Co^{60} nuclei were placed in a magnetic field and cooled to near absolute zero, it was observed that more electrons were emitted in one direction than in the opposite direction.

To help you visualize what nonconservation of parity means from a particle point of view, consider the decay of a π^+ into a μ^+ and a

FIG. 7.3

muon neutrino. Each of these particles has an intrinsic spin and these spins must always be in opposite directions, since the pion spin is zero. The neutrinos observed in nature are always "left-handed," that is, they advance in the same sense as a left-handed screw. But when this experiment is viewed in a mirror, as illustrated in Fig. 7.3, one sees a right-handed neutrino. Since right-handed neutrinos are not found in nature, the mirror image of this experiment cannot occur. Thus pion decay provides an example of non-conservation of parity.

The mirror image of this experiment may be made to represent reality if the mirror images of the muon and neutrino are also charge-conjugated: the μ^+ is replaced by its antiparticle, the μ^-, and the neutrino is exchanged for an antineutrino. Since antineutrinos are "right-handed," our PC mirror has reflected a physically possible interaction.

In 1965 a violation of PC symmetry was observed in the decay processes of K^0 mesons and anti-K^0 mesons. The violation of PC symmetry means that, if the PCT theorem is correct, time reversal applied to the impossible result of a PC-violating experiment would change it into a possible event. But applying time reversal alone to a possible event would change it into an impossible event.

7.5 TIME REVERSAL

The principle of time-reversal symmetry implies that a motion picture of a real physical event, when run backward, also represents a possible event. Movies that are occasionally run backward for comic effect certainly represent series of events that are highly unlikely. As an example, think of a motion picture of a baseball game run in reverse. One might see a ball pop out of a fielder's glove, fly through the air and strike a bat, then rebound to the pitcher's hand. These events, however improbable they might appear, violate no physical laws.

The observation of time reversal in our universe has been the subject of a certain amount of fascinating speculation. A whole galaxy for which time is reversed would have electromagnetic radiation flowing *into* rather than away from it, from our point of view. Conversation with a person for whom time is reversed would certainly be frustrating, for what you consider is the past is his future, and vice versa. The White Queen in *Through the Looking Glass* said, "It's a poor sort of memory that only works backwards!" Yet that's the only way our memories work. Yet memory for a time-reversed man proceeds in the opposite direction from your memory.

In 1922 F. Scott Fitzgerald wrote a story, "The Curious Case of Benjamin Button," in which he described a man for whom time ran backward. Benjamin Button was born in 1860 as a 70-year-old man. He married at age 50, fought in the Spanish-American War, and at age 20 entered a university. In 1914 he graduated, at age 16, and enlisted in the army. Since the records indicated that he had served as a lieutenant colonel in the Spanish-American War, he was promoted to brigadier general. But when he reported for duty as an adolescent boy, he was sent home. Finally he returned to his crib and died.

Benjamin Button was aging backward, but the other events in his life proceeded in a normal manner. A more complete example of time reversal was depicted by Lewis Carroll in "Sylvie and Bruno." There he described a time-reversed meal, during which a person would bring his fork to his mouth, remove a piece of meat, place it on his plate, and "uncut" it with a knife, once more joining it to a larger portion of meat.

These examples of time reversal on a human scale lead to seemingly nonsensical situations. But what happens on the scale of elementary particles? The American physicist Richard Feynman has suggested that certain fundamental processes may be explained on the particle level in terms of time reversal. He noted that the relativistically correct expression for the total energy of a particle is

$$E^2 = m^2c^4 + p^2c^2.$$

Thus

$$E = \pm\sqrt{m^2c^4 + p^2c^2}.$$

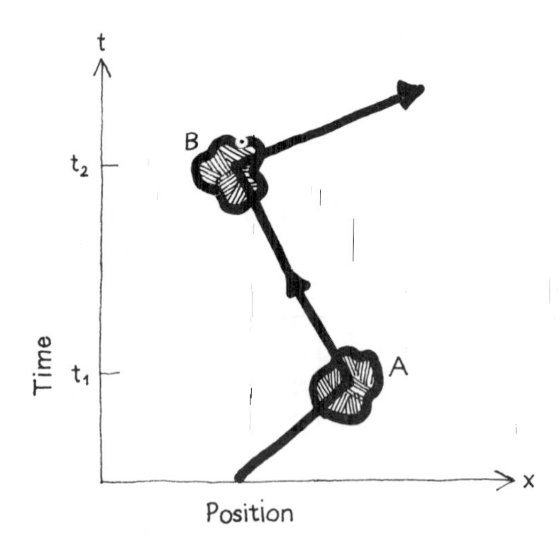

FIG. 7.4

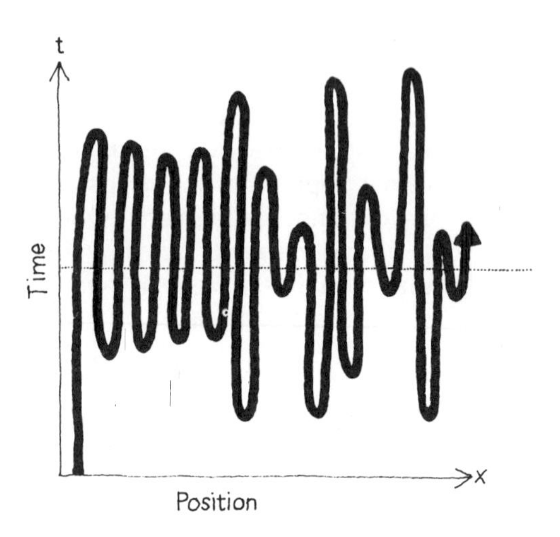

FIG. 7.6

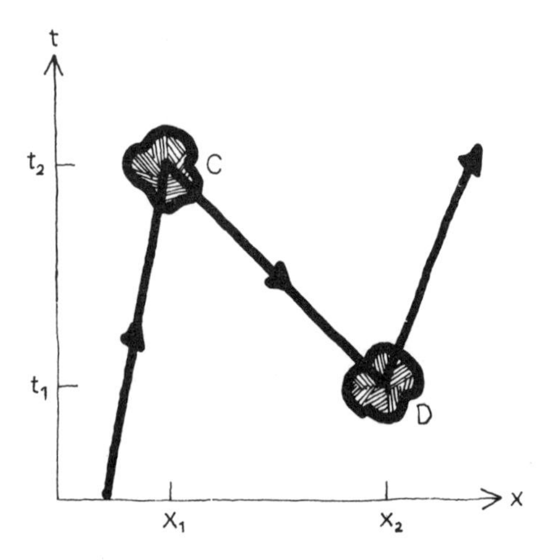

FIG. 7.5

(In these equations, m represents the particle's mass, p its momentum, and E its total energy.) This means that, at least mathematically, both positive and negative solutions to our equations are possible. One may therefore think of particles and antiparticles moving in opposite directions in time.

To illustrate this concept, think first of some ordinary particle interaction. In Fig. 7.4 a particle, say an electron, is scattered at points A and B but, of course, always moves forward in time. In Fig. 7.5, another scattering process is represented. An electron has an interaction at C at time t_2, with a positron leaving C, moving back in time to t_1 until it reaches D, from which an electron emerges. If you were viewing this event from your perspective in time, you would see only one particle, an electron, before time t_1. At t_1 you would observe an electron-positron pair created at D, after which the positron would encounter the original electron at C and annihilation would follow. So long as you view time moving in only one direction, you must use the pair production/annihilation explanation for this series of events. However, if time reversal is allowed, this interaction involves the simple scattering of only one particle. The whimsical suggestion has also been made that the reason all electrons have the same mass and electric charge is that there is only one electron in the universe. The "proof" that this could be the case is presented in Fig. 7.6.

7.6 WHICH PARTICLES ARE FUNDAMENTAL?

The enormously complex situation we have been describing in this chapter presents one with some 200 "particles." Which of these particles are truly fundamental? Which particles are the real building blocks of the universe, particles of which everything else is made?

It has been suggested that many of the so-called particles live so short a time that they have no right to be called particles at all. These short-lived particles, which have lifetimes of only about 10^{-23} second, are known as resonant particles or *resonances*. Their lifetimes are so short that during the same period of time light can travel only the distance across a nucleus. Still other particles are thought to be but excited states of still more fundamental particles.

The history of the search for the most basic constituent parts of matter has led always to the discovery of smaller and smaller entities. Large-scale matter was found to be composed of molecules, which in turn are made of atoms, which were found to contain still smaller nuclei orbited by electrons. This process of discovery has been compared to peeling an onion. Are nucleons and electrons themselves composed of still more fundamental particles? One hypothesis suggests that nucleons are composed of particles known as *quarks*, which carry electrical charges of $-\frac{1}{3}$ and $\frac{2}{3}$ the electron charge.

Quarks (the name was taken from James Joyce's "Three Quarks for Mister Mark" in *Finnegan's Wake*) would exist in threes—if they exist at all. One quark would carry a positive electric charge which would be two-thirds the magnitude of the electron charge, while the other two would be negative, each having one-third the charge of an electron. In this scheme a proton would be made up of two of the $+\frac{2}{3}$ charge quarks and one of $-\frac{1}{3}$ charge, for a net charge of +1 unit. A neutron would consist of a $+\frac{2}{3}$ charge and two $-\frac{1}{3}$ charge quarks, for a net electric charge of zero. The other properties of quarks could similarly be combined to produce all the baryons and mesons.

Although the quark concept is an appealing one from the standpoint of helping to bring about some much-needed orderliness to the elementary-particle situation, unfortunately there is no convincing evidence that quarks exist. In 1969 there was a report that a few particles *might* have been seen which had an electric charge two-thirds that of a proton, but the investigation continues.

7.7 THE ATOM SMASHERS

The whole process of determining the composition of the nucleus and finding particles that are possibly still more fundamental has to be done indirectly. The technique used is analogous to determining how a building is built by firing bullets at it and observing what comes off. From the pattern in which the bullets are scattered one may first learn something of its size, and from an examination of what is knocked out of it one may learn something about what is inside. In particle physics, the "bullets" are highly energetic particles that have been accelerated in particle accelerators, or come from cosmic radiation. The higher the energy of these bullets, the more particles are produced and the more elementary the particles that are produced.

After the development of the first particle accelerator—by J. D. Cockcroft and E. T. S. Walton in England in 1932—the physicist was no longer limited to nuclear bombardment energies no higher than those he could obtain from decaying nuclei. A great many different types of accelerator have now been made, but basically all of them are involved in accelerating charged particles through electrical potential differences. If an electron is accelerated through a potential difference of one volt, that electron acquires an energy of one *electron volt*. The energies acquired by the particles of current-day machines are measured in terms of millions of electron volts, abbreviated *MeV*, or billions of electron volts, abbreviated *GeV*. (Not everybody agrees on whether a billion is one thousand million—as we see it—or one million million. The prefix G has been universally adopted to mean 10^9.)

During the last few years, particle physicists have sought to build particle accelerators having higher and higher energies, in the hope that many of the mysteries surrounding the fundamental particles might be resolved. Thus far, more new questions have been raised than have been answered. At present the world's most energetic particle accelerator is at Serpukhov in the U.S.S.R. This machine has an energy of 76 GeV. Another large accelerator is the 28-GeV facility located near Geneva, operated by a group of European nations at a center known as CERN.

The largest accelerator presently operative in the U.S. is the 33-GeV machine at the Brookhaven National Laboratory (on Long Island, near New York city). A still larger accelerator under construction at Batavia, Illinois—near Chicago—will operate at energies as high as 500 GeV when completed.

If quarks exist, they will not be easy to observe because of the extremely high energies required to blast them loose from the particles they form. One fond hope associated with these giant atom smashers (a misnomer, since atoms were "smashed" at far lower energies; these are really nuclei—and even particle—smashers) is that with their help the mystery of the most fundamental structure of all matter will be solved. However, if one were to judge from our past experiences, something altogether unimagined may be discovered. Such is the nature of the advance of physics.

FOR MORE INFORMATION

Victor Guillemin, *The Story of Quantum Mechanics,* New York: Scribner's, 1968. Contains a very readable discussion of elementary particles, symmetry groups, and quarks.

Eugene P. Wigner, "Violations of Symmetry in Physics," *Scientific American,* December 1965, page 28.

Martin Gardner, "Can Time Go Backward?" *Scientific American,* January 1967, page 98.

Martin Gardner, *The Ambidextrous Universe,* New York: Basic Books, 1964.

Philip Morrison, "The Overthrow of Parity," *Scientific American,* April 1957.

Geoffrey Burbidge and Fred Hoyle, "Anti-Matter," *Scientific American,* April 1958.

QUESTIONS

1. When and by whom was the first subatomic particle discovered? The second?
2. How many particles were known when deBroglie and Schrödinger did their work?
3. Since the particles in the nucleus all carry positive electric charge, why doesn't the nucleus fly apart due to electrical repulsion?
4. Who solved the problem of the "nuclear glue"?
5. What is meant by a charge-independent force?
6. What is meant by an exchange force?
7. What is the exchange particle of the nuclear force?
8. What was the particle first (incorrectly) identified as the Yukawa particle?
9. What is the Yukawa particle?
10. With what phenomenon is the "weak" force identified?
11. What particle is named in Italian?
12. What is the family name of particles that interact via the strong interaction?
13. Which particles interact via the weak interaction?
14. What is the difference between the two types of hadron?
15. What are the two types of baryon?
16. Why are protons and neutrons both described as nucleons?
17. Can a lepton be "strange"?
18. What property do baryons and leptons have in common that mesons do not have?
19. Is there any particle for which there is no antiparticle?
20. The antiparticle of a proton is an antiproton, yet the antiparticle of an electron is usually known as a positron, not an antielectron. Is there some physical reason for this, or is it a matter of semantics?
21. What was a basic consideration in Dirac's prediction of the existence of the positron?
22. What happens when a particle and its antiparticle get together?
23. What do we mean when we say that a physical quantity is conserved?
24. Why are quantities such as momentum or baryon number useful?
25. What do we mean by parity?
26. What do P, C, and T stand for?
27. When was parity nonconservation first observed?
28. For what type of interactions does one have parity nonconservation?
29. When was PC nonconservation first observed?
30. If PC is not conserved in an interaction, what can you say about T for that process?
31. Is a neutrino left-handed or right-handed?
32. What known physical processes may be described in terms of time reversal?
33. Why are higher and higher energy accelerators needed to do fundamental particle research?

34. What is a quark?

35. How many kinds of quarks are there?

36. What type of mirror is used in Table 7.1?

PROBLEMS

1. We have described parity reversal (P), charge conjugation (C), and time reversal (T) mirrors. What sort of a mirror is the one in Table 7.1?

2. Imagine that radio communications are established with a civilization whose planet orbits a star ten light years distant. Is there any way, without going there, of determining whether their world is made of matter or antimatter?

3. Describe, in as much detail as you can give, a conversation with a time-reversed person.

4. If there are equal amounts of matter and antimatter in the universe, what can you conclude about its distribution?

5. It has been suggested that, in gravitational interactions, antimatter is repelled by matter. If this were true, could antimatter be used in the same way as "cavorite"? (See Problem 1.3.)

6. For which of the following sports could you tell whether or not you were watching the event or its mirror image: baseball, football, tennis, basketball? (Assume that you can't see numbers or lettering on uniforms, and that you don't have prior knowledge of which players are left-handed or right-handed.)

7. Can you think of a sport which would appear normal if it were time-reversed? (Assume that you watch only play in action, not the beginning or ending of play.)

8. The mirror image of a right-handed person is left-handed. Unless reflections about horizontal and vertical axes are different, why isn't a mirror image also upside down?

9. In beta decay, a neutron may decay into a proton plus an electron and an electron antineutrino. You may convince yourself that such things as baryon number, electric charge, and lepton number are conserved in this process. What may a proton decay into?

10. Big particle accelerators cost big money. What justifications can be made for building still bigger ones? What arguments can be made against doing so?

11. Describe the reflection symmetry properties of the following two figures.

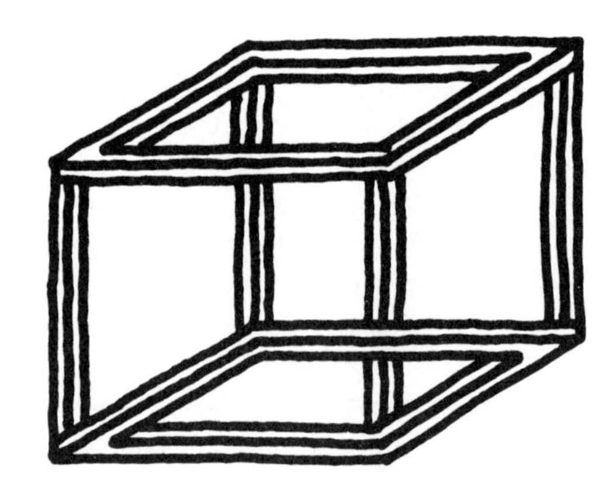

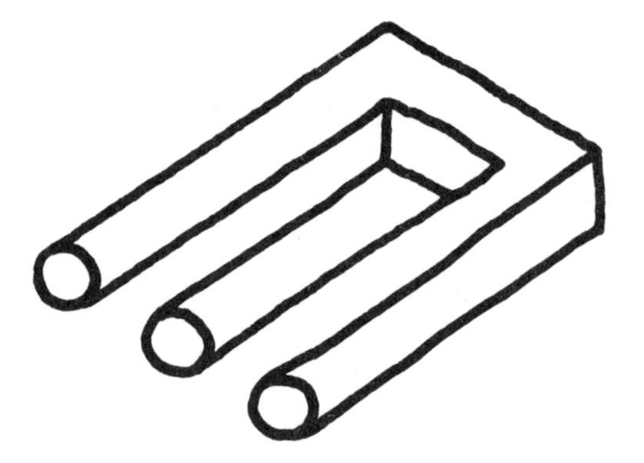

CHAPTER 8 PROPERTIES OF MATERIALS

8.1 ELEMENTARY PARTICLES, ATOMS, AND MOLECULES

In Chapter 7 we were concerned with the "building blocks of the universe": the fundamental, elementary particles. Now let's see how these building blocks are put together to make up all the material in our universe.

We have previously noted that it is impossible, at this point in the development of human knowledge, to say precisely *which* of the nearly 200 identified "particles" are truly fundamental. We have grouped these particles into families, according to their characteristics, but we are unable to determine whether the members of a family must exist separately. It is quite possible that some particles are made up of others that have been put together, while still other particles may be no more than excited states of still more fundamental particles. For example, one could build a proton or a neutron from three quarks.

Whatever *particles* are absolutely fundamental may be used together to build *atoms*. In constructing atoms we shall most certainly require electrons, protons, and neutrons. But we shall also need the field quanta of the forces involved in holding these particles together. The particle associated with the nuclear force, the pion, as well as the particle connected with the weak interaction, the neutrino, are involved in the construction of the nucleus; while the photon—the quantum of the electromagnetic field—is associated with the orbiting electrons. The gravitational force and its particle, the graviton, are only very weakly involved in processes on the atomic scale.

The atoms, thus constructed, would include each of the 92 elements we find in nature, plus whatever synthetic elements we choose to build. We know that there are several isotopes of most of these elements, each of which may have important different physical properties.

From these elements we may then build up all the various *molecules* involved in the more complex structures of matter. Some molecules, such as the inert gases—helium, argon, krypton, etc.—each consist of only a single atom. In several cases a molecule made up of only a single isotopic form of an element is involved in an important structure. For example, the liquid form of He^3 has unique cryogenic properties. Other molecules may be made up of several atoms of a single element. The gaseous forms of nitrogen, oxygen, and hydrogen are made up of diatomic molecules: N_2, O_2, and H_2—two atoms combining to form a stable structure. Still more complex molecules consist of combinations of atoms of more than one element. The hydrocarbons may contain dozens of atoms of hydrogen, carbon, and other elements, while biomolecules such as DNA and RNA may contain thousands of atoms.

8.2 PHASES OF MATERIALS

Materials are normally found in one of three phases—solid, liquid, or gaseous—representing a progression of the material from lower to higher temperatures. In any sample of a substance, it is possible to have more than one phase of the material present. For example, ice always has some water vapor over it, and ice may even exist in equilibrium with both the liquid and vapor present. This precise state of affairs exists at the *triple point* of water.

We have discussed *gases* at some length in connection with planetary atmospheres (Chapter 2). We noted that a gas fills whatever volume it is placed in. The molecules of the gas are in rapid motion, and there is a distribution in speeds of these molecules, dependent on the temperature of the gas.

When sufficiently cooled, a gas may condense into a *liquid*. Of all the liquids we know, water is certainly the most important and interesting. A liquid, in the process of solidifying, normally contracts; water, on the contrary, *expands*. This means, of course, that ice is less dense than water and floats upon it. If water acted like most liquids in the process of solidification, ice would be found at the bottom of lakes and ponds.

Although most liquids become more dense as they cool, reaching maximum density just at the point of solidification, water has its greatest density at about 4°C, which is about 39°F. This means that in the process of being frozen, a body of water is first cooled to 4°C before freezing can take place, and that freezing proceeds from the surface downward. This phenomenon makes it possible for marine life to survive in bodies of water that are subject to being frozen.

There is a great deal of physics connected with fluids, both at rest and in motion, but we shall here mention but one of the most important phenomena: *Archimedes' principle*. This principle, which bears the name of its ancient Greek discoverer, simply states that any object immersed in a fluid will be buoyed up with a force equal to the weight of the fluid the object displaces.

As one illustration of Archimedes' principle, sea water weighs about 64 pounds per cubic foot (Fig. 8.1). Thus an object with a volume of 1 cubic foot will receive an upward, buoyant force of 64 pounds when it is immersed in sea water. Whether this object will sink or float depends on whether it weighs more or less than 64 pounds. Similarly, a 192-pound man will float in sea water if his volume is greater than 3 cubic feet, and he will sink if his volume is less (Fig. 8.2). The usefulness of life preservers and other flotation devices is due to the fact that they increase the effective volume of the user without greatly increasing his effective weight. If one is in a fluid denser than ordinary water, such as the waters of the Great Salt Lake or the Dead Sea, which have high concentrations of dissolved elements,

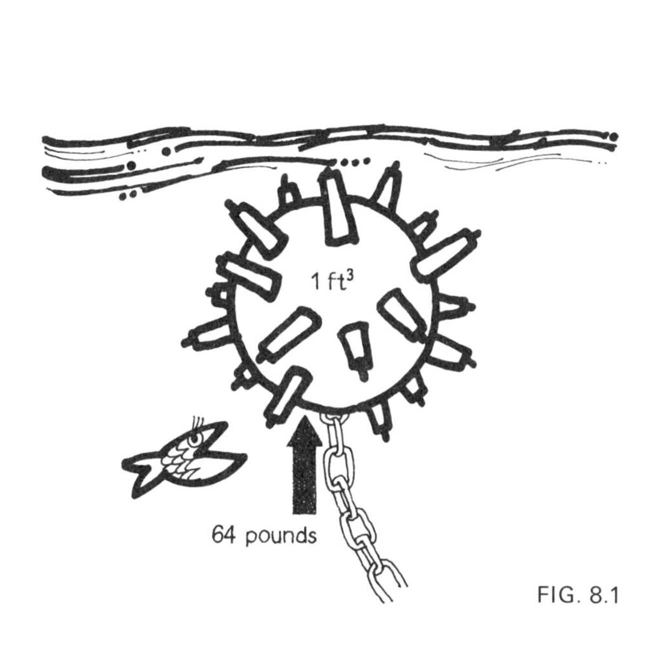

FIG. 8.1

FIG. 8.2

floating is much easier than in ordinary water because the weight of any given volume of this water is greater than the weight of the same volume of ordinary water.

It is in *solids,* however, that the most interesting of the problems of physics are found. Before dealing with the specific properties of certain solids, however, we want to note some important effects that are independent of any particular solid and are concerned only with *scaling factors.*

Other things being equal, the strength of a rope depends on its *cross-sectional area.* If you want a rope that is twice as strong as one that you already have, you don't need to buy one with twice the diameter, for that would mean that you would then have a rope with four times as great an area, and thus a rope four times as strong as the one you had originally. This is also true of any supporting structure, such as the columns or pillars that support a building.

It is sometimes mistakenly thought that structures may be scaled up or down in size by any multiplying factor and still retain their same relative strength. To illustrate that this is so, consider the following simple examples. Suppose that you start with a structure—for simplicity, let's consider a cube supported by circular legs—and scale it upward in size by a factor of 10. Each dimension will then become ten times as great as before. The volume of the cube will be $10 \times 10 \times 10$, or 1000 times as great as before, and its weight—which will vary in direct proportion to its volume—will also be 1000 times as great as it was originally. However, the supports—whose strength depends on their bottom area—will have had their bottom area increased by a factor of only 100, and will therefore be only 100 times as strong as before. As a result, the structure will be loaded by a factor of 10 greater than before, because 1000 times as much weight will now be carried by supports that are only 100 times as strong as before. This sort of calculation may be made in scaling buildings, rockets, animals, and people.

Heating and cooling are also very much involved with scaling factors. Again, for simplicity of calculation, think about a cube of side s, which we may say is a crude model of some animal. The surface area of this cube is equal to $6s^2$, the volume is s^3. The heat produced, in the case of an animal, depends on its volume, while the rate at which heat is lost to the atmosphere depends on the surface area. Thus the heat loss per unit volume depends on a term that varies as $6s^2/s^3$, or $1/s$. If we were to attempt to scale this animal up by a factor of 10, the heat loss would be proportionately lessened, while a division of all dimensions by 10 would increase heat loss by this factor. For this reason, if for no other, an elephant could not operate in the same way if it were the size of a mouse, and vice versa.

Jonathan Swift's Gulliver traveled to Brobdingnag, where the inhabitants were about twelve times our size, and Lilliput, with its miniature people one-twelfth our size. If these creatures were built like scaled-up or scaled-down humans, they would have been plagued by numerous problems we don't have.

Unless the bones of the people of Brobdingnag were made of a much stronger material than ours, they would have been incredibly frail creatures. A human thighbone breaks under about ten times the human weight; thus a Brobdingnagian could have easily broken his leg while merely walking! A column of water as tall as a Brobdingnag man exerts about two atmospheres of pressure at its base. Therefore the problem of blood circulation would have been formidable.

The Lilliputians, on the contrary, would have been quite strong, but they would have been beset with problems of high rates of body heat loss. Unless they had extraordinarily high-calorie-content foods available, Lilliputians would have had to eat almost constantly. Swimming would have been a rugged sport in Lilliput, for a swimmer, in emerging from the water, would have had to lift a film of water comparable to his own weight.

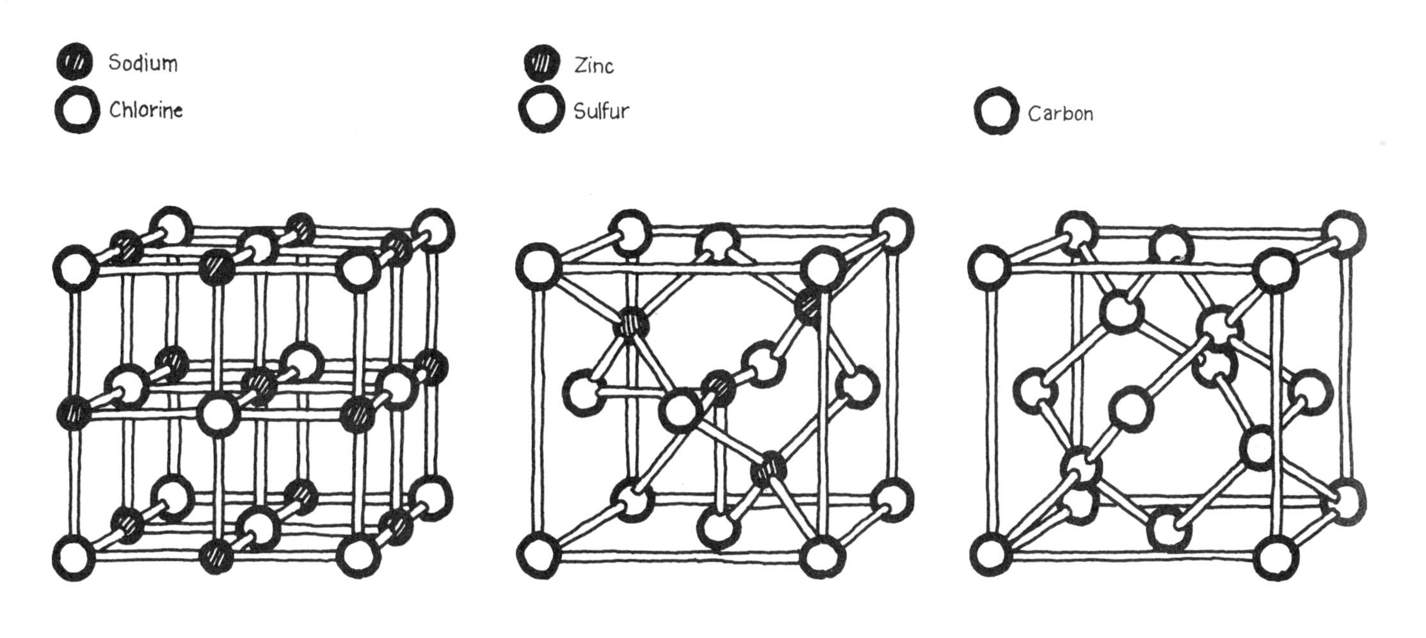

FIG. 8.3 Sodium chloride.

FIG. 8.4 Zinc sulfide.

FIG. 8.5 Diamond.

8.3 STRUCTURES OF SOLIDS

Although many materials exist in amorphous, disordered forms, by far the most interesting structures are those that possess an ordered structural pattern. Materials whose atoms are arranged in regular, periodic structural patterns are known as crystals. Properties of crystals have been the object of intense study by physicists working in many different areas of research.

The very simplest sort of crystal structure is the cubic lattice arrangement. The sodium chloride (table salt) structure, illustrated in Fig. 8.3, has this form. This is simply a rectangular lattice with the sodium and chlorine atoms alternating in rows in whatever direction one proceeds. Note that each sodium atom has six chlorine atoms for nearest neighbors, while each chlorine atom has six sodium atoms for nearest neighbors.

More complex structures are represented in Figs. 8.4 and 8.5. One of the most complex structures known is that of the diamond lattice. Diamond is made up of carbon atoms, just as graphite is made up of carbon atoms, but when these atoms are arranged in the diamond-lattice structure, they form the extremely hard substance that is familiar to us as diamond.

For many of our studies, the important things about lattice structures are the periodicities and symmetries of the lattice. These properties may not be fully appreciated unless one can view a three-dimensional lattice model of the particular crystal. In dealing with the property of electrical conductivity, we shall be particularly concerned with the matter of periodicity.

8.4 ELECTRICAL CONDUCTIVITY

Quantities such as heat, light, sound, and electrical currents may be conducted through materials. By far the most important conductivity property with which we shall be concerned is that of electrical conductivity. We normally think of materials as being either electrical conductors or electrical insulators, for most materials have electrical properties near these extremes. The range of electrical conductivity is perhaps the widest of any known physical property, with good conductors conducting by as many as 32 orders of magnitude better than good insulators.

Metals and Insulators

We shall begin our discussion of electrical conductivity by considering metals, since they are, in general, good electrical conductors. In a metal structure such as silver or copper or gold, there are electrons, known as *conduction electrons,* that are not bound to any one lattice atom, but are free to range throughout the material. We often make useful calculations by considering the structure as if it contained an *electron gas.* The other electrons in the material, the ones bound tightly to lattice atoms, need not concern us at the moment.

If a piece of metal is placed in a closed circuit and an electrical potential difference is applied to the ends of the metal, there is a flow of electric current—electrons will move—along the closed circuit. If you think of the electrons as an electron gas, you may wonder how it is possible for these electrons to go very far in the structure without colliding with the atoms at the lattice sites. One might think that they would follow random paths like ordinary gas molecules would in moving through a gas, colliding with lattice atoms and other electrons frequently. But here the business of quantum mechanics and lattice periodicities comes in. You will remember that electrons possess wavelike properties. If the lattice is perfectly regular, it is possible for an electron to move easily through the lattice if the electron wavelength is a multiple of the distance between lattice sites. In the hypothetical perfect lattice, an electron could move throughout the material without encountering any resistance.

Any real lattice is always less than perfect, and anything that upsets the perfect periodicity of the lattice will impede the electron flow. Two sources of resistance to electron flow through materials arise from *lattice imperfections* and *thermal effects*. Lattice imperfections may consist of impurity atoms that have been substituted into the lattice, missing lattice atoms, atoms placed in the lattice in the wrong places, or a variety of other faults. These effects are independent of temperature, and the part of the electrical resistance in a metal that is due to lattice defects is not temperature dependent. Thermal motion (or vibration) of the atoms bound to lattice sites also presents resistance to the passage of electrons through a material. The higher the temperature, the greater the thermal motion; thus the component of electrical resistance due

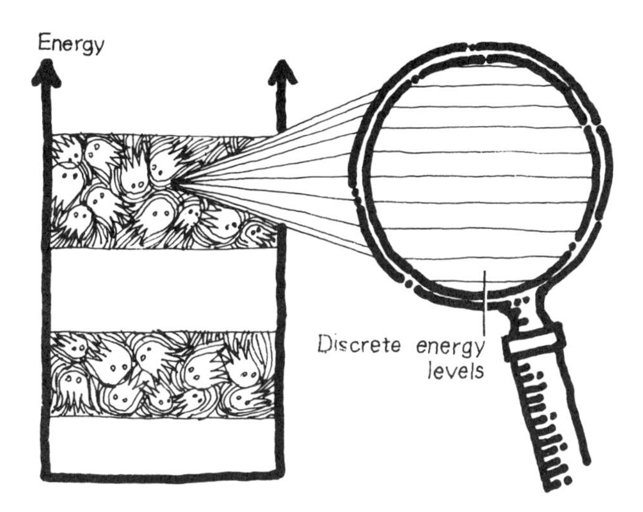

FIG. 8.6 Energy bands.

to these vibrations is temperature dependent. In a metal, the electrical resistance therefore increases as the temperature rises.

From our discussion of the Pauli exclusion principle, we know that each of the electrons in this electron gas in our conductor has four quantum numbers associated with it, and no two electrons in our system are permitted to have precisely the same set of quantum numbers. Each of these sets of quantum numbers represents a possible electron energy level. In some cases two or more different sets of quantum numbers may correspond to levels having the same energy values, though in most cases they differ. These energy levels are very close together, so close together, in fact, that they are masked by thermal effects and we cannot meaningfully distinguish between adjacent levels in a material with a great number of electrons present. This is the case for even very small samples of any material. Therefore we must consider these many discrete energy levels as constituting a broad *energy band* (Fig. 8.6). Due to the periodicities of the lattice, some energy levels are not available to the electrons. These are known as forbidden energy levels, and they make up *forbidden energy bands.*

In any sample of material, we may represent the locations of all the electrons in the material in terms of the energy bands occupied by these electrons. It is in the context of the filling of energy bands that we may distinguish between good and poor conductors of electricity.

The movement of electrons through a material makes up an electric current. To have a current flow, one must first have electrons available to be a part of that current. This alone is not a sufficient condition for electrical conductivity, since it is also necessary for an electron that is a part of a current flow to have an energy level available into which it may go when its

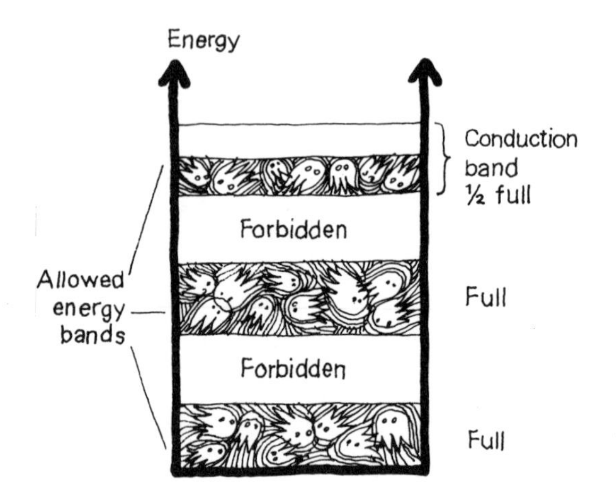

FIG. 8.7 Metal.

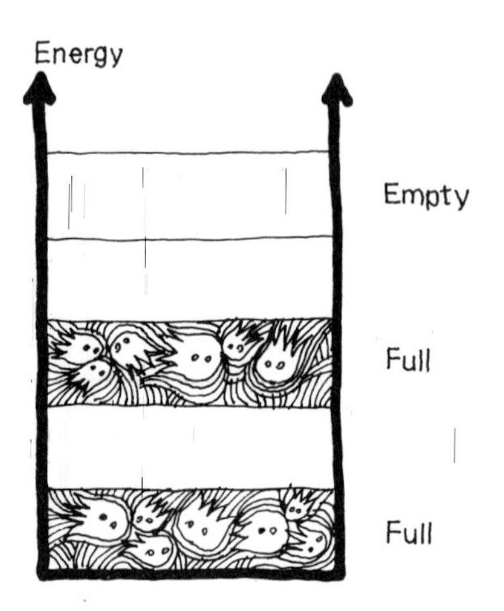

FIG. 8.8 Insulator.

energy level is increased by the potential difference that must be applied to initiate the current flow. It follows from the exclusion principle that you cannot place an electron in an energy level that is already filled, nor can an electron exist anywhere other than in an allowed energy level.

Now we are ready to understand why there is such a difference between the electrical conductivities of metals and insulators. Figure 8.7 shows the electron energy-band structure of a typical metal. All the allowed energy bands below the top one are filled, and we are not concerned with them in the conduction process. None of the electrons in these lower bands can be raised into higher energy levels in the same band, because these levels are all filled, so the only way one of these electrons could participate in the conduction process would be for it to receive enough energy to be raised all the way to the conduction band. Such an amount of energy is not normally available. But note that the conduction band is half filled (and half empty). Thus each of the electrons in this band has a place to go at only a slightly higher energy level, and all the electrons in the conduction band are thus available to take part in the conduction process. We have therefore satisfied the two necessary criteria for electrical conduction: there are plenty of electrons available and there are energy levels into which they may go.

Figure 8.8 shows the band structure of a typical insulator. Note that all the bands that are occupied by electrons are filled, and that there is a considerable gap between the last filled band and the first empty one. For electrical conduction to take place in this material, it is necessary for one to apply a potential great enough to raise electrons from

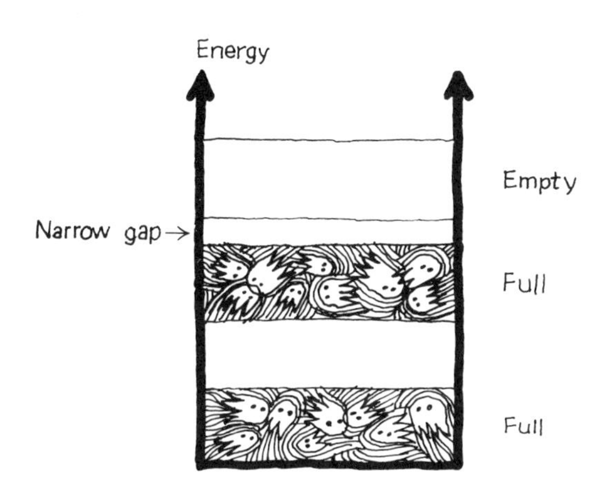

FIG. 8.9 Intrinsic semiconductor.

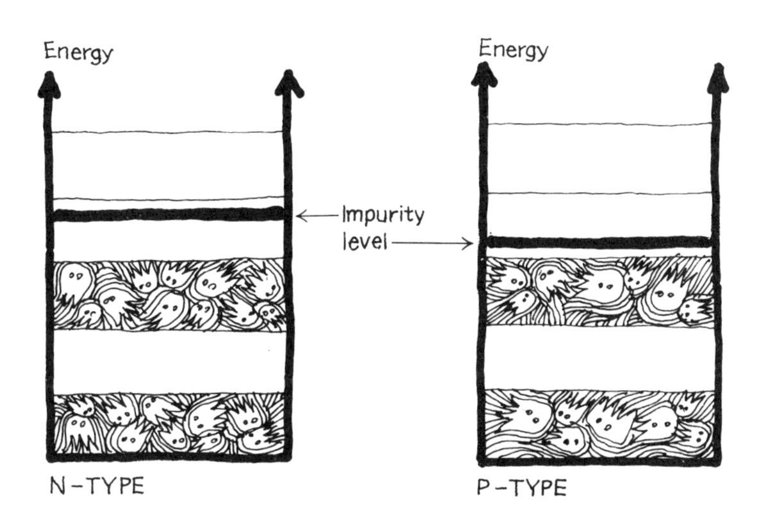

FIG. 8.10 Impurity-type semiconductors.

the highest filled band into the first empty (conduction) band. Since a great deal of energy is required for this, the electrical resistance of this material is high.

Semiconductors

Not all materials fall at the two extremes of being either excellent conductors or very poor conductors of electricity. Between these extremes there is a range of materials which are classified as *semiconductors.*

There are two basic types of semiconductor material: those that are *intrinsic* semiconductors and those that are made semiconductors by the addition of *impurities*. Figure 8.9 shows the band structure of an intrinsic semiconductor. Note that—just as in a good insulator—the last energy band containing electrons is filled and the conduction band is empty. But note also that the gap between these bands is not so great as it is in an insulator. This gap may be so small that the application of a low potential to the material can raise electrons into the conduction band, or electrons may even be thermally raised across the energy gap. Thus the material may be made to conduct electricity, though it will not be as good a conductor as a metal because fewer electrons will be taking part in the conduction process.

Intrinsic semiconductors are rare, however, and for practical applications impurity-type semiconductors are almost always used. In an impurity-type semiconductor, the idea is to add a convenient additional energy level made possible by the presence of the impurity material. An example of the way in which this is done is illustrated in Fig. 8.10. Consider a lattice of a material such as silicon or germanium.

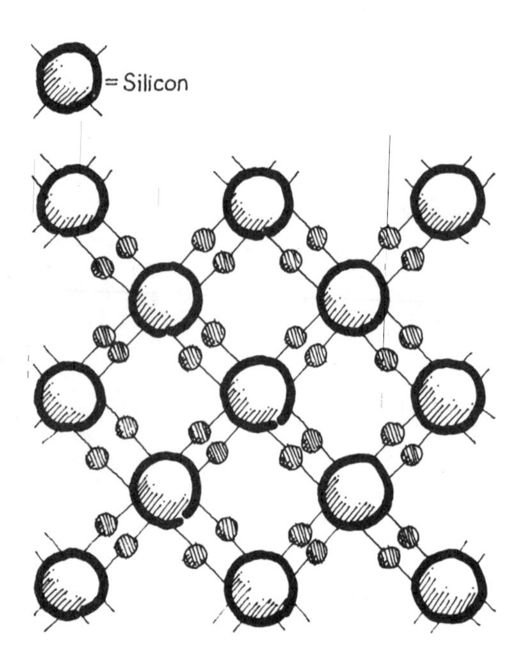

FIG. 8.11 Normal silicon lattice.

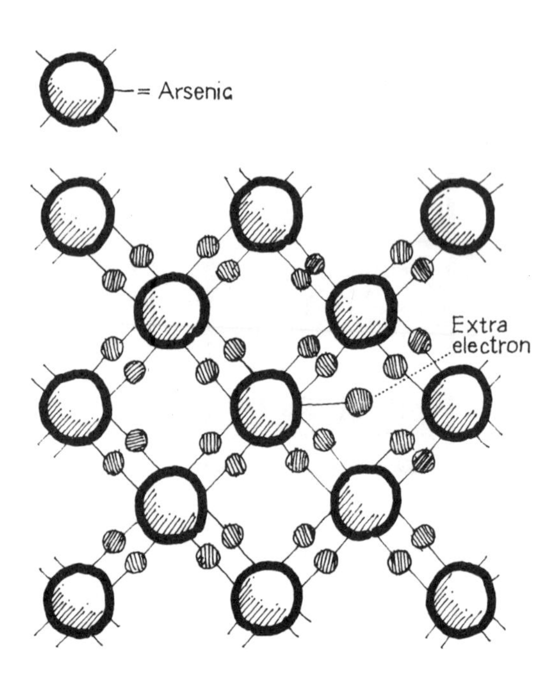

FIG. 8.12 Donor, or n-type, semiconductor.

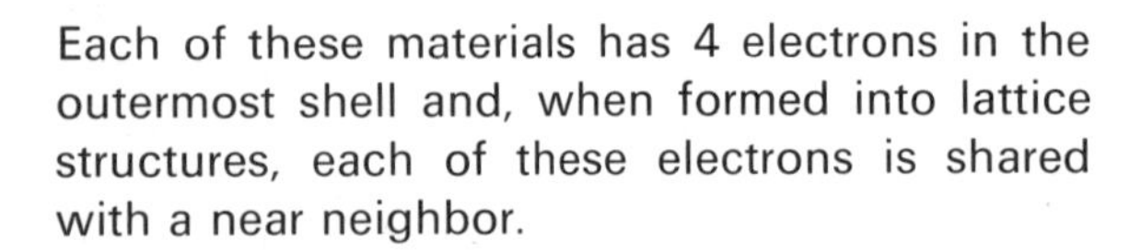

Each of these materials has 4 electrons in the outermost shell and, when formed into lattice structures, each of these electrons is shared with a near neighbor.

Figure 8.11 shows a normal silicon lattice. If one of the silicon atoms were to be replaced in the lattice by an atom of an impurity material, such as arsenic, the situation shown in Fig. 8.12 would exist. Arsenic has 5 electrons in its outermost shell, one more than silicon. Thus there is an extra electron available for each arsenic atom substituted into the silicon lattice. The type of semiconductor thus formed, with extra electrons available for conduction, is known as a *donor*-type semiconductor. The negatively charged electrons are the charge carriers in the conduction process; thus the material is known as an *n-type* semiconductor. It is also possible to construct a semiconductor using an impurity element containing one electron fewer than the material of the lattice. Gallium has only 3 electrons in its outer shell, and the substitution of a gallium atom in the silicon lattice provides a situation like that shown in Fig. 8.13. There is now a vacancy where there should be an electron; we call this a *hole*. An electron from an adjacent pair of atoms may hop over and fill this vacancy, but this in turn will leave a hole at the site from which the electron came. If another electron fills the newly formed hole, it may be said

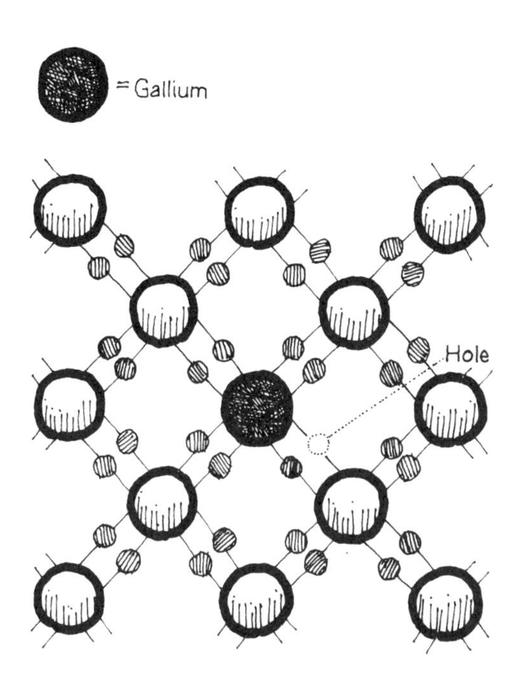

FIG. 8.13 Acceptor, or p-type, semiconductor.

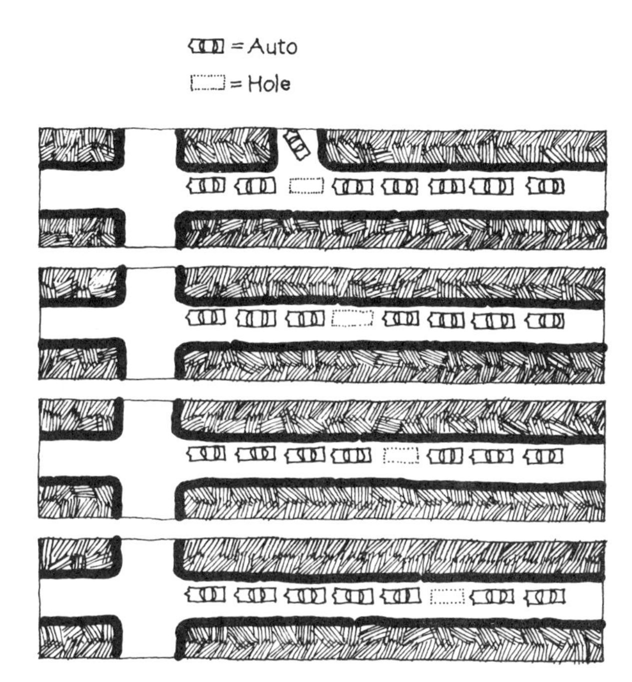

FIG. 8.14

that electrons are moving over to fill the hole, or it may be said with equal validity that the hole is moving through the material.

We may present an analogy to the movement of holes from a scene that you might see when looking out a window from a building overlooking a busy street. Imagine a line of cars stopped at a traffic signal (Fig. 8.14). Suddenly one of the cars in the line makes a turn into a parking garage, leaving a gap in the line of cars. The vehicle immediately behind the one that turned out will pull up to fill the place, but there will now be a gap behind *it*. The next car in the line will move forward to fill the new gap, and so on. You might describe what has happened by saying that each of the cars behind the one that turned out has pulled up one place. But you might also say that a *hole* was created in the line and that the hole moved backward.

Once you are willing to consider a hole as a physical reality, we may describe the process of electrical conduction by holes as easily as that by electrons. Of course the electrons are the things that are moving, but the departure of an electron from a spot—the lessening of the negative charge there—produces precisely the same effect as the arrival of a positive charge. It is simpler, in semiconducting materials of this type, to think of the holes as being the charge

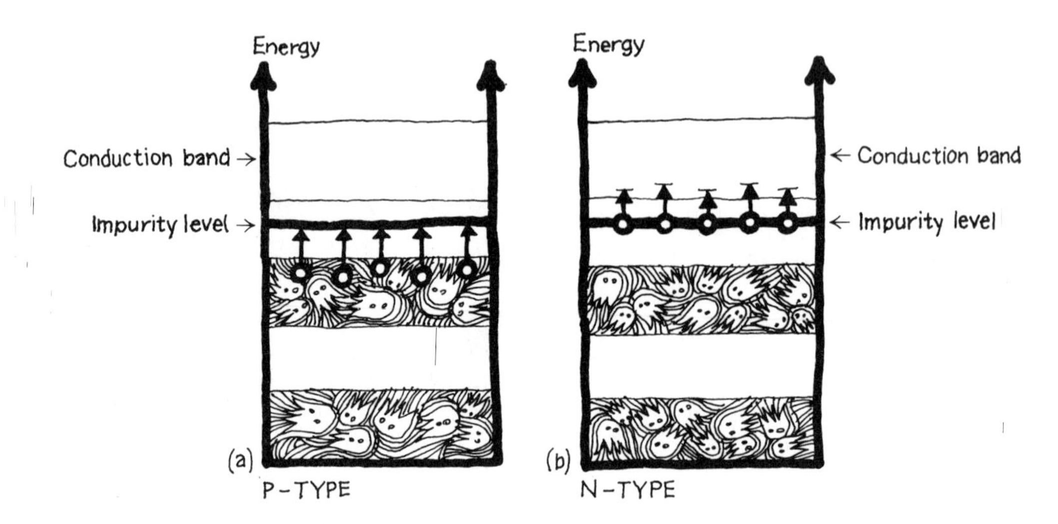

FIG. 8.15(a) Electrons thermally raised into conduction band form impurity level. (b) Holes created as electrons are thermally raised into impurity level.

carriers. This type of semiconductor, known as an *acceptor* type, is called a *p-type* semiconductor.

In both n- and p-type semiconductors, the impurity levels are near the conduction bands; thus mere thermal excitation may move electrons (or holes) into these bands (Fig. 8.15). The more thermal energy available, the more charge carriers available for conduction. This means that, in semiconducting materials, the conductivity increases with temperature and the electrical resistance decreases as the temperature rises—a behavior opposite to that of metals.

The amount of impurity that these impurity-type semiconductors require is very small. In a typical material, one impurity atom to each 10^{10} atoms of the base material is sufficient. To make materials with impurities this carefully controlled, it is first necessary to refine ultra-pure materials, then add impurities as required.

8.5 USES OF SEMICONDUCTOR MATERIALS

So far we have described a type of material that is neither a very good conductor nor a very good insulator. Since you know that semiconductor materials are responsible for a great revolution in electronics, you may be wondering what all the clamor has been about. The answer is that the two varieties of semiconducting materials, n-type and p-type, when used together, may form devices that will do just about anything a vacuum tube can do, and do it in ways a vacuum tube cannot hope

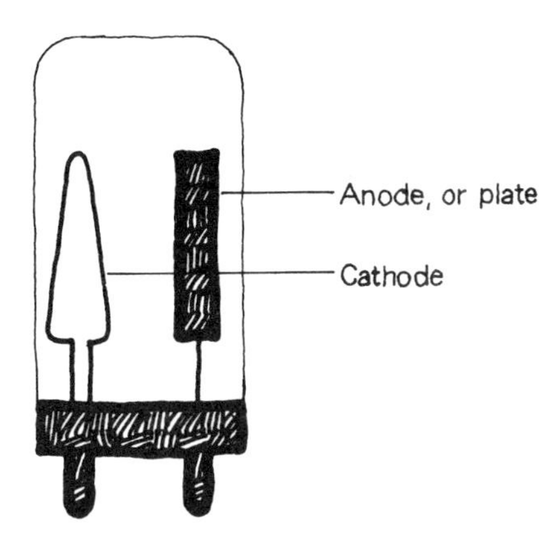

FIG. 8.16 Vacuum-tube diode.

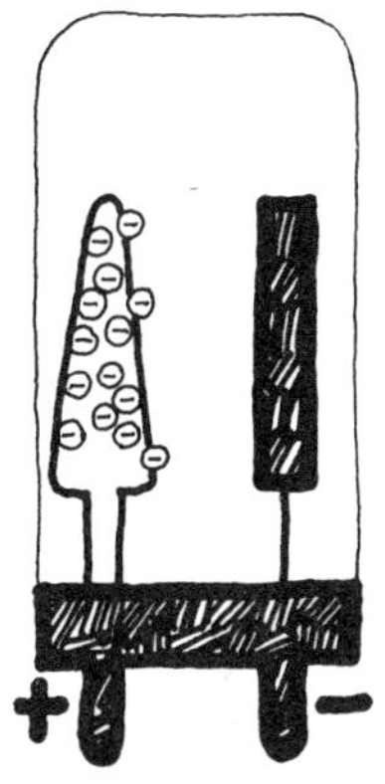

FIG. 8.17 Diode operation.

to match. To explain the uses of these materials, we first need to understand what is actually done by vacuum tubes.

Diodes

The simplest type of vacuum tube is known as a *diode*, since it has just two elements: the emitter of electrons, the *cathode*, and the element that receives electrons, the *anode* or *plate* (Fig. 8.16). In a typical vacuum tube the cathode and plate are enclosed in an evacuated tube (hence the name *vacuum* tube) and the cathode is heated. (This is where the dull red glow of a tube originates.)

When the cathode is heated in the vacuum, electrons are "boiled off," as they receive enough energy to escape the metal cathode. The plate of the tube, if given a positive bias, attracts these electrons, and there is a flow of current from the cathode to the plate through the tube (Fig. 8.17). The tube may now be turned on or off by the application of a signal to the plate. When the plate is made negative with respect to the cathode, the electrons are repelled by it and the flow of current ceases. Biasing the plate positively again restores the electron flow. Thus one of the uses of a diode is as an electronic switch. (The name *electron valve* has been applied to these devices.)

Another use of diodes involves the conversion of an alternating current (ac) to a direct current (dc). This process is called *rectification*. If an alternating signal is applied to the plate of a diode, only half the signal is carried, since the plate is negative half the

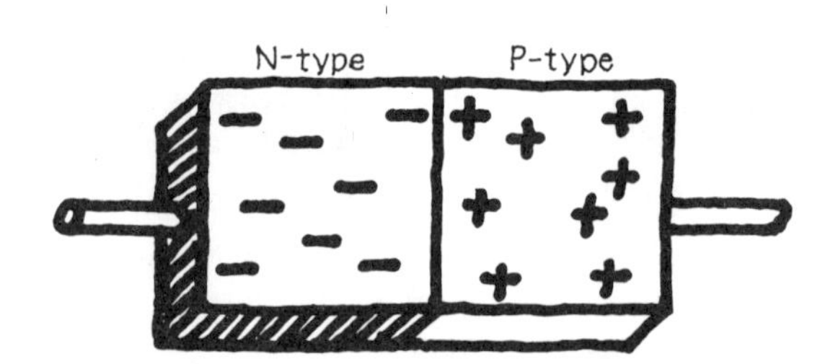

FIG. 8.18 Semiconductor diode.

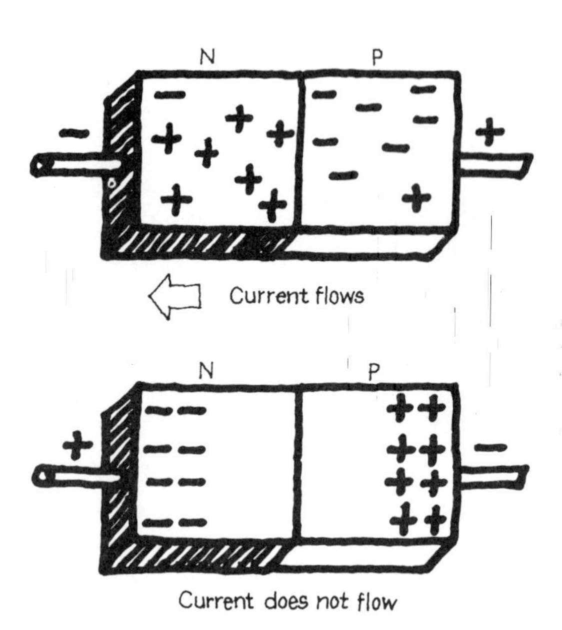

FIG. 8.19 Semiconductor diode operation.

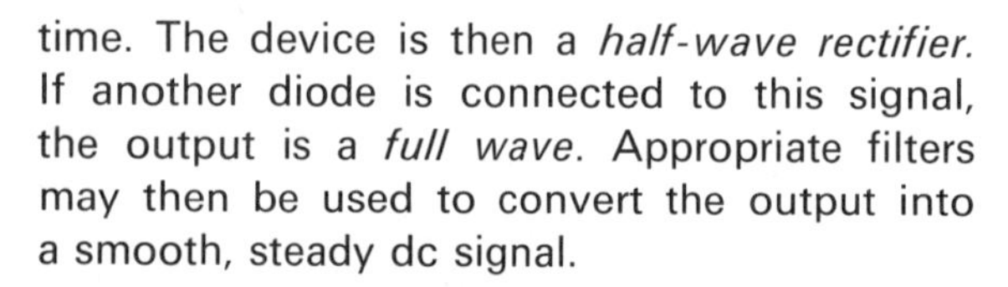

time. The device is then a *half-wave rectifier.* If another diode is connected to this signal, the output is a *full wave.* Appropriate filters may then be used to convert the output into a smooth, steady dc signal.

A semiconducting diode (Fig. 8.18) consists simply of an n-type semiconducting material joined to a p-type material. When the p-type material is connected to a positive potential, the resistance of the diode is small and current may easily flow through it (Fig. 8.19). When the p-type material is biased negatively, current does not flow. Thus the combination of the two types of semiconductor material acts as an electronic valve, just as the vacuum-tube diode does. Hence *semiconductor diodes* may be used for this purpose in place of vacuum tubes.

Triodes

The addition of a screen *grid* to a vacuum-tube diode between the cathode and the plate makes it possible for a *triode* (3-element) vacuum tube to be used for amplification of a signal (Fig. 8.20). Note that the grid is placed much nearer to the cathode than to the plate. If the plate is left positive (and the tube is in the conducting mode) the tube may now be turned on or off by imposing a small negative potential on the grid. When the grid is negative, electrons will not travel to the plate despite the fact that the plate may be at a high positive potential. The grid is made in the form of a screen mesh so that electrons can pass through it. Thus the grid may be used to control the current that flows through the tube. The output

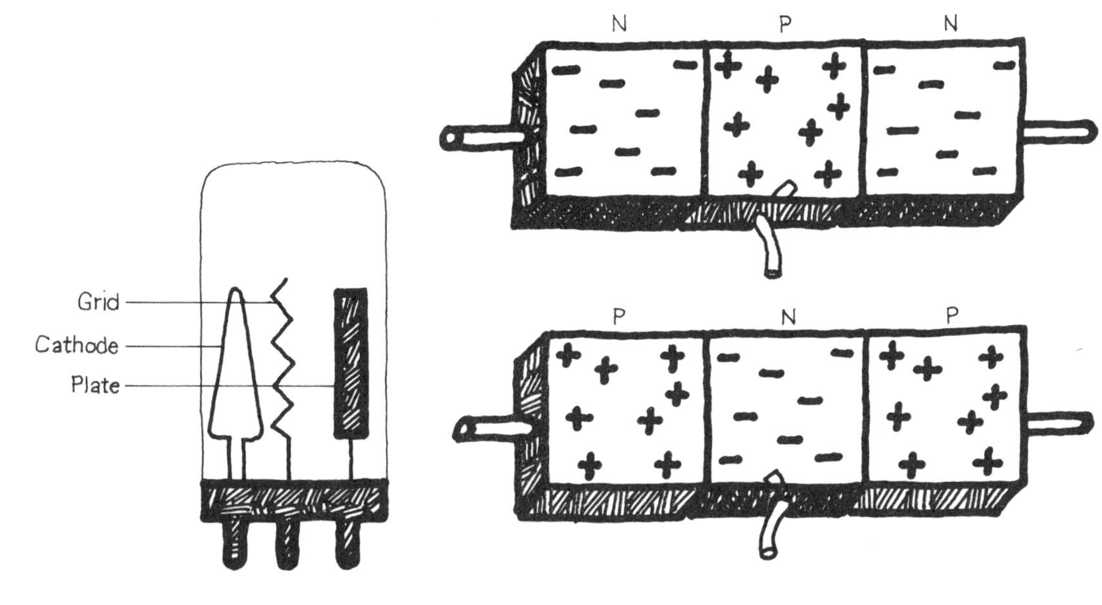

FIG. 8.21 Semiconductor triodes or transistors.

FIG. 8.20 Vacuum-tube triode.

signal at the plate has the same pattern as the input at the grid, but a much greater amplitude. Thus the triode may *amplify* a signal that is fed into it.

To construct a semiconducting triode, one may make either an n-p-n or a p-n-p sandwich. In either case, the semiconducting triode is normally called by its more familiar name, the *transistor* (Fig. 8.21).

Amorphous Conductor Devices

A very new type of solid-state electronic device that is still in the development stage is the *amorphous conductor,* which consists of a mixture of elements in an amorphous (i.e., not ordered, like a crystal) state, placed between two conductors. These devices have the remarkable property of not conducting electricity until a certain voltage threshold across them is reached; then they become very good conductors. Amorphous conductors may therefore act as very sensitive, high-speed electrical switches having many potential applications in computer circuitry. It has also been suggested that these devices may make it possible to construct solid-state TV picture screens. Such screens might be made having virtually any desired viewing area, but no more than an inch or so thick.

8.6 COMPARISON OF SOLID-STATE AND VACUUM-TUBE DEVICES

The widespread replacement of vacuum tubes by solid-state devices may be easily understood when the operations of the two types of device are compared. A vacuum tube has a *limited lifetime:* the filament may burn out, in time gas may leak into the tube, or other parts may fail. In applications in which reliable operation is mandatory, schedules calling for the routine replacement of vacuum tubes after a fixed number of hours of operation are necessary. In contrast, the expected lifetime of solid-state devices is measured in at least the tens of years, so long as rated voltages are not exceeded.

Vacuum tubes consume a great deal more *power* than solid-state devices. Much of this power is used in heating filaments, which of course transistors do not have. Not only does a tube drain power from its power supply, but the heat generated by this power dissipation often must also be removed. The first generation of electronic computers, which used large numbers of vacuum tubes, required high-capacity air-conditioning systems to maintain reasonable operating temperatures in the units.

The transistor is inherently more *rugged* than a vacuum tube. A tube, when subjected to mechanical shock, may fail even if the glass envelope does not break.

Tube heaters require *time to warm up* before the device will operate; solid-state electronic devices are "instant on." Because solid-state devices are physically much smaller than vacuum tubes and require lower operating voltages and power levels, it is possible to take advantage of circuit miniaturization to reduce the *size* of the electronic circuits.

8.7 SOLID-STATE CIRCUITRY

Solid-state electronic devices such as semiconductor diodes and transistors by themselves would have made possible many advances in electronic circuits and applications. But their usefulness has been greatly increased by the fact that designers of electronic circuits have been able to take advantage of their small size and low power drain. Complex electronic circuitry, such as that required by computers, has been reduced in size by several orders of magnitude, and still further size reductions in circuits will be forthcoming.

Printed Circuits

The devices that go into making up a useful electronic circuit must be electrically connected in order for the circuit to work. Until a few years ago, the connections used were usually lengths of conducting wire (Fig. 8.22). Circuits constructed in this way have the advantage of being relatively easy to repair or modify, but they are not very rugged mechanically

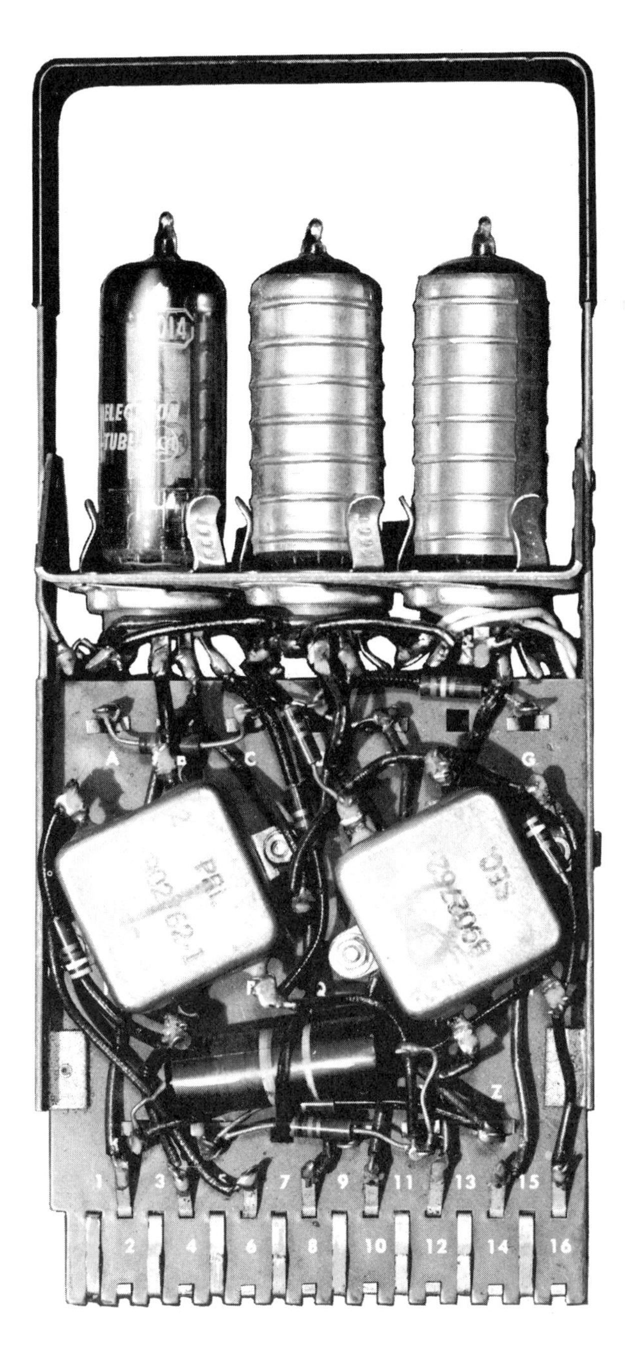

FIG. 8.22 A typical wired circuit, such as was used before the advent of the transistor. (Courtesy of RCA)

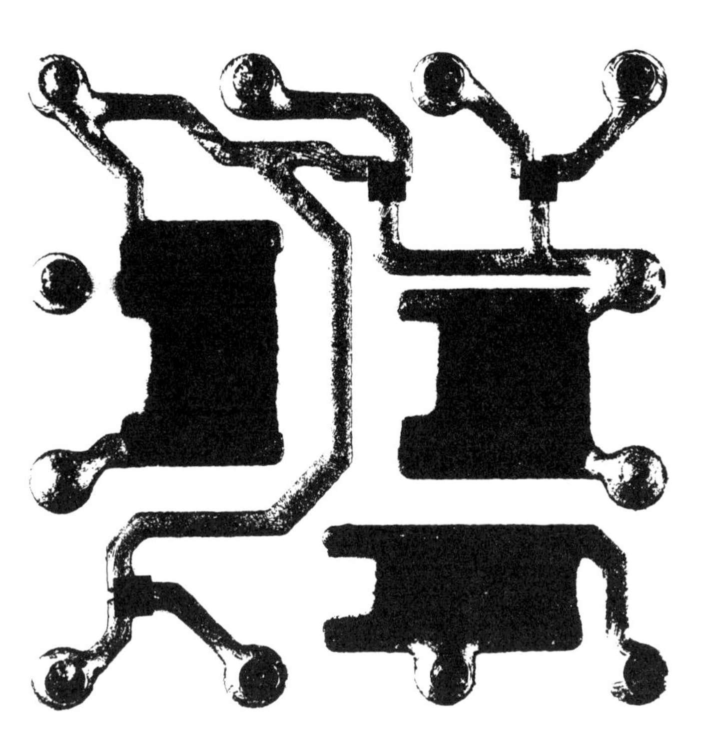

FIG. 8.23 Printed circuit board, much enlarged. It is actually about the size of a typewriter key. (Courtesy of IBM)

and do not occupy the smallest available space. A first step in making devices more compact consisted of utilizing printed circuits.

A printed circuit (Fig. 8.23) consists of an arrangement of conducting paths placed on an insulating board by a photo-etching process. The components used in the circuit are mounted on this board with their terminals soldered to the printed circuit conductors. Since there are no dangling wires or components suspended in space, this type of circuit is less vulnerable to mechanical damage than a conventionally wired circuit. Further, since the components normally are all mounted on one side of the circuit board, making connections to the printed circuit may be done by dipping the board into a solder bath. Thus the manufacturing operation may be automated and the chances for wiring errors during production are reduced.

When a manufacturer is packaging components which use printed-circuit boards, these boards may be attached to the rest of the circuit by demountable connectors. This factor speeds repair operations when the unit is in service. Note that vacuum tubes can be and often are included in printed-circuit applications. However, boards using only solid-state devices are more compact and more rugged than those incorporating vacuum tubes.

Modular Circuits

A modular circuit really consists of any group of components that has been packaged into a sealed module. The device is then treated like a "black box" in operation: it is plugged in, and, if it doesn't work properly, the whole unit is replaced (Fig. 8.24). Modules using only solid-state electronic devices may be made very compact and mechanically rugged. Servicing is reduced to an operation no more complicated than replacing fuses: a semi-skilled person simply plugs in new modules until the piece of equipment works again. Of course, repair costs can be measurably greater in modular-component uses, if, for example, a $20.00 module (or even a $200 module) must be replaced because a 10-cent resistor inside it has failed. On the other hand, if your TV set consisted of modular circuits only, instead of calling a repairman when it failed, you might stop on the way home from work and put down a deposit on one of each module, take them home and plug them in until the set worked. The unused, good modules would be returned and you would pay to replace only the module that had actually failed.

FIG. 8.24 Front and back views of plug-in modular circuit boards. (Courtesy of RCA)

Integrated Circuits

An *integrated circuit* is made up of solid-state electronic devices that have been formed directly on—and are a part of—the supporting structure itself (Fig. 8.25). For example, connections and parts of circuit components

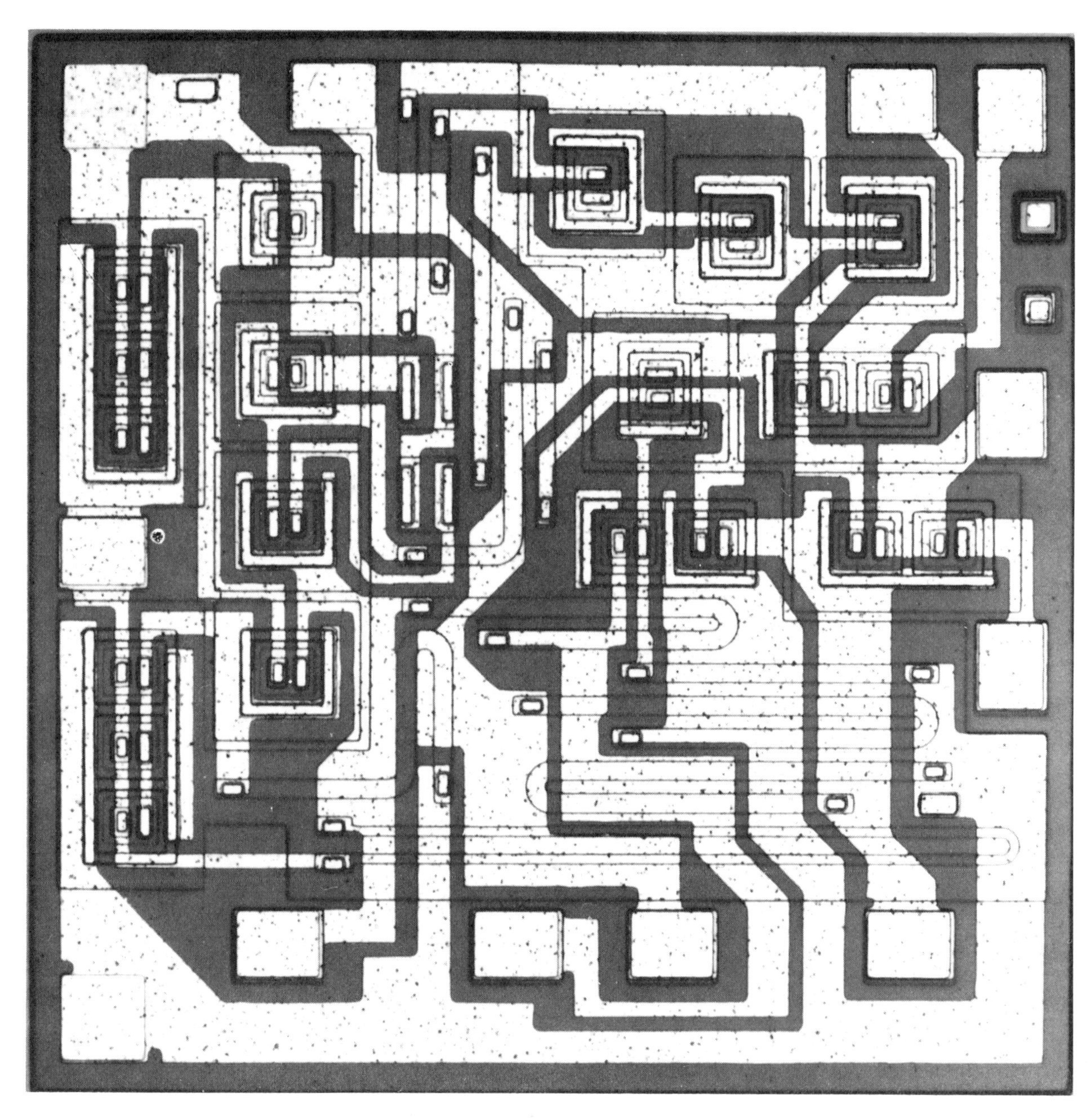

FIG. 8.25 An integrated circuit, consisting of n-p-n transistors, diodes, and resistors. (Courtesy of Motorola)

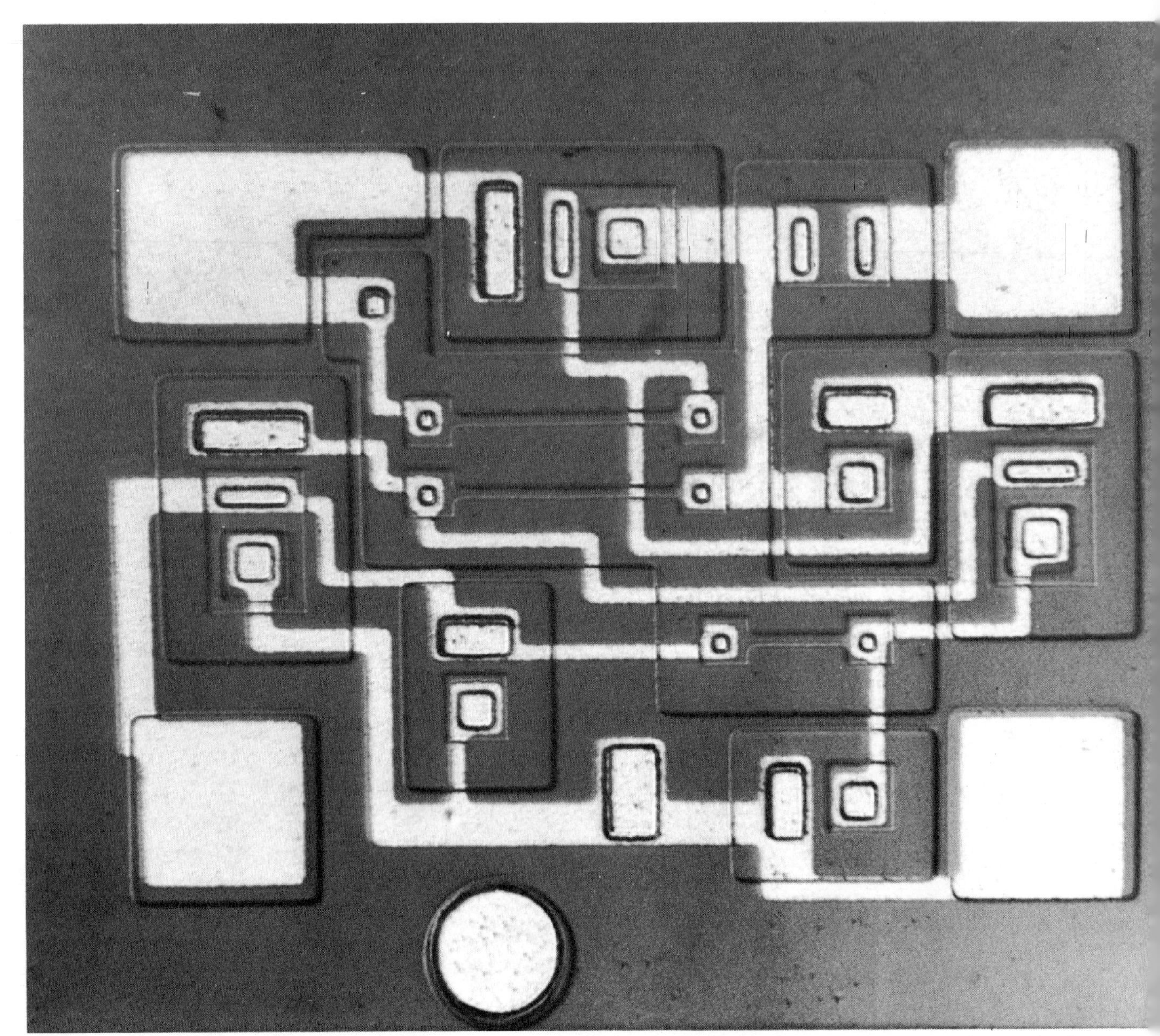

FIG. 8.26(a) A typical micro-integrated circuit, much enlarged. The density and actual size of such circuits is illustrated by part (b) on opposite page. (Courtesy of P. R. Mallory Co.)

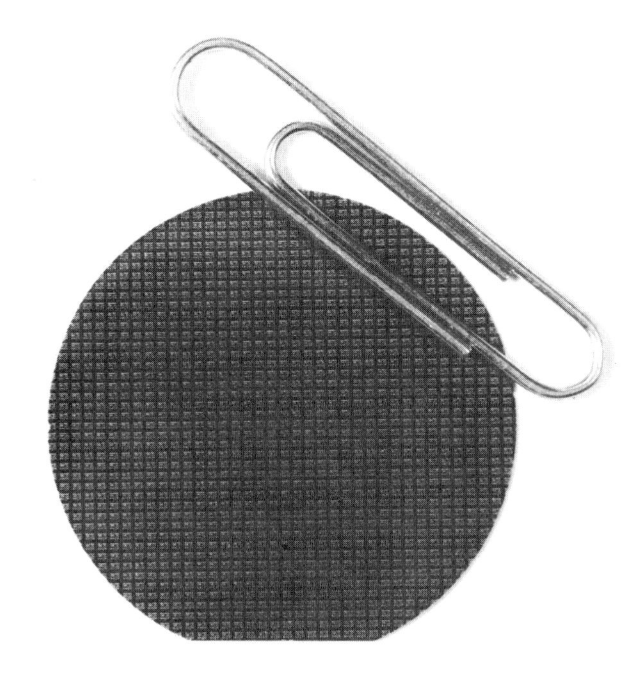

FIG. 8.26(b) A one-inch slice of silicon, containing some 1000 integrated circuits, such as the one shown on the opposite page. The paper clip in the photograph is included to give an idea of size. Due to the extremely small patterns made possible by photolithography, a very large number of components can be fabricated simultaneously in a very small area. (Courtesy of P. R. Mallory Co.)

may be formed by thin films of materials deposited directly onto the surface of a thin silicon wafer, which itself provides the semiconductor material needed for transistors and diodes. What we are saying is that instead of separately making electronic circuit components and bringing them together and joining them into a circuit, one forms the circuit components—transistors, diodes, resistors, capacitors, etc.—directly on a small chip of semiconductor material. The typical micro-integrated circuit may have 25 or more transistors and associated circuit components on a silicon wafer only one-tenth of an inch on a side (Fig. 8.26). We have by no means reached the limits of possible size reduction, and still smaller circuits will undoubtedly be made. Using circuits like these, it would be possible to put all the electronic devices in your TV set (picture tube excluded) into a small matchbox, and the circuitry needed for an AM radio would cover only the head of a pin.

8.8 THERMOELECTRICITY: AN ALMOST OVERLOOKED DISCOVERY

In 1820 Hans Christian Oersted discovered that an electric current flowing through a wire causes the needle of a compass placed nearby to be deflected. We now know that Oersted had simply observed the fact that any flow of current produces a magnetic field around the conductor of the current, and it was this magnetic field that deflected his compass needle.

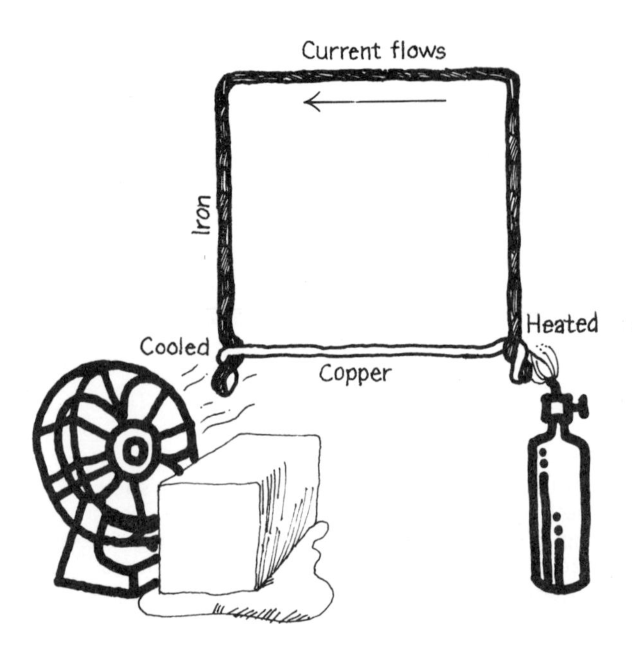

FIG. 8.27 The Seebeck effect.

Oersted's discovery was still not fully understood when, a year later, Thomas Johann Seebeck found that if he made a circuit of two dissimilar electrical conductors and heated one junction while cooling the other, a magnet placed nearby would be similarly deflected. Seebeck was convinced that magnetism had something to do with this heating and cooling, and eventually tried to explain the existence of the earth's magnetic field in terms of the temperature difference between the poles and the equator. Not only was Seebeck's explanation wrong, but he was able to keep others from discovering the true meaning of his discovery for a number of years.

Seebeck was the first person to report the observation of the *thermoelectric effect.* Figure 8.27 illustrates the basic idea involved. If one makes up a circuit of two different metals, such as iron and copper, and creates a temperature difference between the two bimetallic junctions, a current will flow if the circuit is closed. This phenomenon is known as the Seebeck effect, after its discoverer. In 1834, Jean Charles Peltier found that if a current is made to flow through a junction made from two different metals, the junction will be heated or cooled depending on the direction of the current flow. Thus if an external source (such as a battery) causes a current to flow in a circuit like the one Seebeck used, one of the junctions will be heated and the other cooled (Fig. 8.28).

The Seebeck effect has been widely used as a temperature-sensing device: one junction is maintained at some constant, reference temperature—usually in a bath of ice and water—

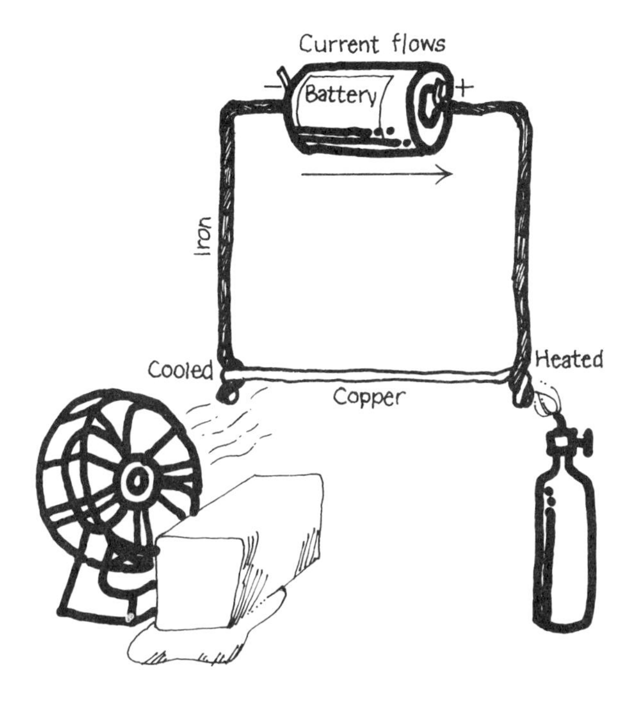

FIG. 8.28 The Peltier effect.

while the other junction is placed in contact with an object whose temperature is to be measured. The difference in potential across the circuit may be directly related to the difference in the temperatures of the hot and cold junctions. A circuit set up in this way is known as a *thermocouple thermometer.*

But the most important potential application of the Seebeck and Peltier effects was not recognized for almost 100 years. This is particularly surprising when we consider that the device used by Seebeck had a thermal efficiency of about 3% (that is, it would convert 3% of the thermal energy incident on it into electrical energy), which compared favorably with the efficiency of the steam engines used in his day. These men had discovered means of generating electricity and of heating and cooling objects, but their nineteenth-century discoverers did not develop them.

Modern Thermoelectric Devices

As long as one sticks to metals for making thermocouples, there is a definite built-in limitation to their efficiency. The very material properties that make metals good conductors of electricity also make them good conductors of heat. Thus heat flows along the wires of a thermocouple along with the electrical current, and the efficiency of the device is limited. To minimize the heat flow, one can use small wires, but this limits the amount of current that may be carried.

However, it is interesting to find that Seebeck himself, in experimenting with a variety of materials for his thermocouple junctions, used some of the materials we know today as semiconductors. The advantage of using a semiconductor instead of a metal in making a thermoelectric device is simply that semiconductors—although they are fair conductors of electricity—are normally poor conductors of heat.

If one takes metal wire and heats one end, there will be an immediate thermally activated flow of electrons from the hot end to the cold end of the wire. The absence of the electrons from the hot end and their build-up at the cold end creates a potential difference between the two ends of the wire, but this condition does not last long. First, electrons are attracted back to the hot end of the wire, since it is, briefly, electrically positive. Second, because the wire conducts heat well, the cold end of the wire is quickly warmed and other electrons are driven back along the wire to very nearly restore charge equilibrium to the conductor. However, the heating of a semiconductor produces a much greater difference of potential between the ends of the material. There are not so many conduction electrons to be involved in the semiconducting material, and there is less heat transfer to the cooler end of the material. Thus a potential difference of a fraction of a volt may be set up between the ends of the semiconducting material.

The type of semiconductor described above is an n-type material, with electron charge carriers just as in a metal. In an n-type semiconductor, the hot end of the wire becomes positive, the cold end negative. In a p-type semiconductor, the opposite polarity is obtained: the charge carriers are the holes, which move in the direction opposite from the electrons. In a p-

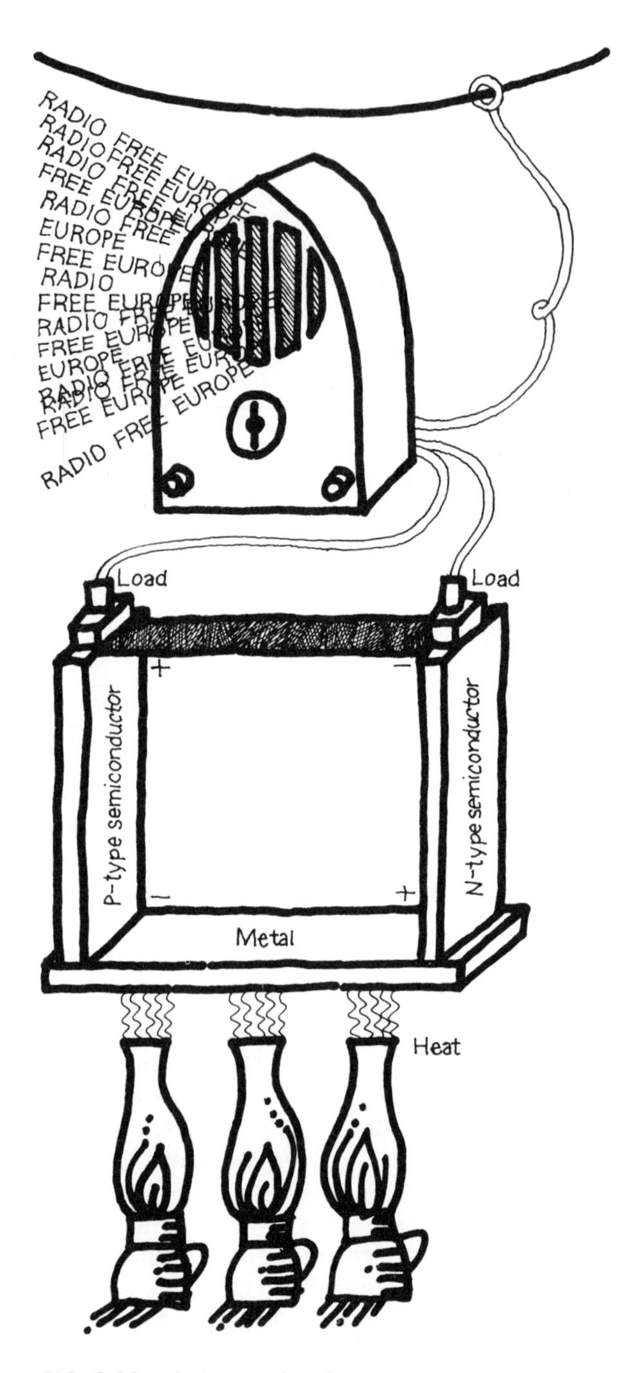

FIG. 8.29 A thermoelectric generator.

type semiconductor, the hot end becomes negative and the cold end positive.

To form an effective thermoelectric device, one has only to connect a piece of n-type and a piece of p-type semiconductor to a strip of a metal (Fig. 8.29). When the metal is heated, a difference in potential is set up at the ends of the semiconductors. These two terminals serve the same purpose as the two terminals of a battery. One has formed a device that, when heated, is an electric generator.

Generators of this sort may have a thermal efficiency of about ten percent today. This is less than that of steam-turbine electric-generating devices, but thermoelectric generators have been under active development for a much shorter period of time. Each cell of the device described typically produces a potential difference of about one-tenth of a volt. Thus many such cells are connected to form useful generating devices.

The Peltier effect—the reverse of the Seebeck effect—may be used to produce either heating or cooling by causing a current to flow through the device. Thus, depending on the direction of flow of the current, the device may serve either as a refrigerator or a heater (Fig. 8.30).

Applications of Thermoelectric Devices

The thermoelectric devices we have described have a number of inherent advantages over conventional devices that produce the same effect. Solid-state thermoelectric devices are small, extremely rugged, have no moving parts, and have very long lifetimes. Present applications of these devices include using them as small electric generators and as compact refrigerator units.

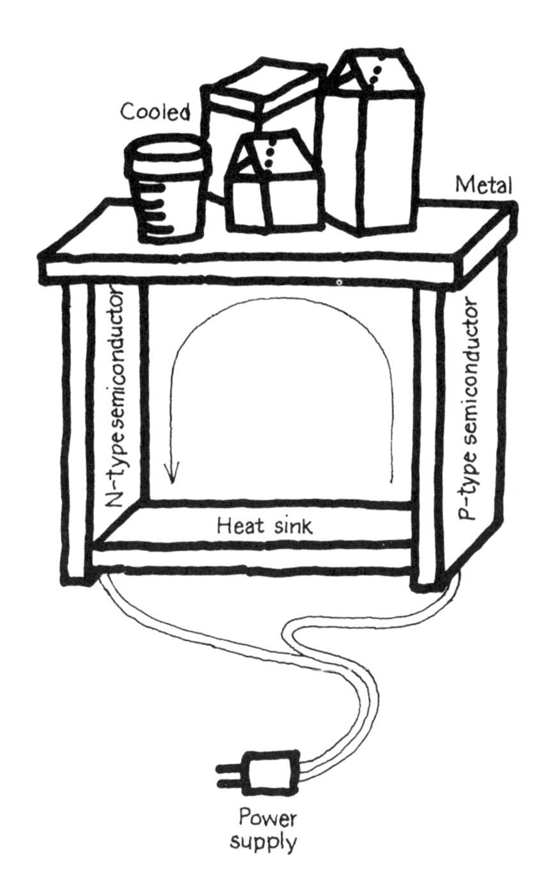

FIG. 8.30 A refrigerator.

The passive seismometers left on the face of the moon by the Apollo missions were powered during the long lunar nights by small thermoelectric generators that were heated by radioactive material. For years, peasant homes in Siberia have used thermoelectric generators, heated by kerosene lamps, to operate radio receivers. The Peltier effect has also been used to produce small portable refrigerators and miniature heating devices.

The potential applications of thermoelectric units are almost unlimited. Sources of heat—particularly waste heat—abound. One of the major problems of many industries is cooling materials or equipment. Some of the heat that is now exhausted into the air or water, to serve only as sources of thermal pollution, could be used to generate electricity. An automobile has a heat engine which must be cooled by air or water, and the heat produced must be dissipated into the atmosphere. Some of this now-wasted energy might also be used. In addition, most automobiles can slow down from a speed of 60 miles per hour to a standstill in about 4 or 5 seconds; yet to accelerate that same car from a standstill to 60 miles per hour in the same 4 or 5 seconds would require about a 600-horsepower engine. That's the amount of power you use in stopping your car. Where does this power go? Most of it is dissipated by your brakes, providing another source of thermal energy that could be converted into electricity.

Nuclear reactors produce energy in the form of heat. All nuclear power plants now operating and planned use this heat to produce steam and use conventional turbine electric generators. Future plants may be designed to produce electricity directly by means of thermoelectric devices.

Of course the greatest source of thermal energy available on earth is solar radiation. A battery of thermoelectric cells supplied with the energy that falls on an area about 100 miles square would supply the entire present electrical power needs of the earth's population. While such a facility is hardly practical on this scale, it is important to realize that some of those areas of the world most in need of economical sources of electric power are in regions in which there is abundant sunshine. If we could use the sun to supply power, it would not be impractical to consider thermoelectric generating plants large enough to provide power for irrigation of regions that are presently deserts.

FOR MORE INFORMATION

Alan Holden, *The Nature of Solids,* New York: Columbia University Press, 1965. A nonmathematical treatment of solids, their structures and properties.

Gregory H. Wannier, "The Nature of Solids," *Scientific American,* December 1952.

Abram F. Joffe, "The Revival of Thermoelectricity," *Scientific American,* November 1958, page 31.

J. B. S. Haldane, "On Being the Right Size," *The World of Mathematics,* Volume 2, New York: Simon and Schuster, 1956.

F. G. Heath, "Large-Scale Integration in Electronics," *Scientific American,* February 1970, page 22. A survey of microelectronics and a description of the latest applications of ultra-miniaturized circuitry.

William Hume-Rothery, *Electrons, Atoms, Metals and Alloys,* New York: Dover, 1963. This book is written as a dialog between an older metallurgist and a young scientist. Although a certain amount of mathematics is used, readers without an extensive background in mathematics will still find most of the conversations quite readable. (In paperback.)

QUESTIONS

1. List the hierarchy of structures of materials from the largest to the smallest assemblages known.
2. In what phases may materials exist?
3. How does the volume of a given sample change with the phase in which the material appears?
4. At what temperature does water have its greatest density?
5. Contrast the difference between what happens to water and what happens to most other liquids as they are cooled and then solidified.
6. State Archimedes' principle.
7. Why is it easier to float in sea water than in fresh water?
8. An object will not float on the ocean's surface. If it is released at a point where the sea is 10,000 feet deep, will it sink all the way to the bottom?
9. Compare the relative strengths of the people' of Lilliput and Brobdingnag.
10. How does the relative rate of body heat loss compare for small and large animals?
11. What physical property has the widest known range of values?
12. Do all the electrons in a metal take part in the conduction process?
13. What is an electron gas?
14. What gives rise to electrical resistance in metals?
15. How does the resistance of a metal change as its temperature is increased?
16. What does the Pauli principle have to do with the conduction electrons of a metal?
17. What is a forbidden energy band?
18. What causes the great difference of electrical conductivity between metals and insulators?
19. How does a semiconductor differ from a conductor? From an insulator?
20. Distinguish between intrinsic and impurity semiconductors.
21. Sketch an n-type semiconductor and describe its operation. Repeat this for a p-type semiconductor.
22. What determines whether an impurity will make an n- or p-type semiconductor?
23. What is a hole?
24. Why does a hole carry a positive charge?
25. What are the necessary components of a vacuum-tube diode? A semiconducting diode?
26. What added component turns a vacuum-tube diode into a triode? What is added to make a semiconducting triode?
27. What name is usually used for semiconducting triodes? What can they be used for?
28. What do the letters n-p-n and p-n-p signify?
29. What are some possible uses of amorphous semiconductor devices?
30. Compare solid-state and vacuum-tube devices, listing the advantages each type of component has over the other.
31. Compare hand-wired, printed, modular, and integrated circuits. Tell how each type of circuit differs from the others, and list the advantages each type has over the others.
32. What is the Seebeck effect? The Peltier effect?
33. How old is the science of thermoelectricity?
34. What are some uses for thermoelectric devices?
35. What characteristic of the thermoelectric generator makes it particularly suited for use with nuclear power sources?
36. List some inherent advantages thermoelectric generators have over conventional ones.
37. What produces the greatest power output in your car?

PROBLEMS

1. What would you expect to find different about our world if water were to contract on freezing?

2. A ship has a displacement of ten tons in sea water. What is its displacement in fresh water?

3. You can see a balloon inside a car just ahead of you in traffic. As that car turns to the right, you see the balloon displaced to the right also—to the *inside* of the turn. What is responsible for this balloon's strange behavior?

4. An elephant has been described as a mouse that was designed by a committee. If an elephant were a scaled-up mouse, what problems or advantages would it have?

5. Comment on the statement, "If a flea were the size of a man, he could jump a hundred feet into the air."

6. Graphite and diamonds are both made only of carbon atoms. Account for the great differences in the physical properties of the two materials.

7. If there were no exclusion principle, what sort of electrical-conductivity properties would you expect to find for materials?

8. Imagine that you are chief executive officer for General Widget, Inc. Your line of frammistats now uses vacuum tubes and hand wiring, but the boys in the lab are anxious to go to solid-state modular circuitry. What *questions* would you want answered before you give the go-ahead to this move?

9. From what you now know about scale and volume/surface relationships, what optimum design features should be incorporated into buildings to be built in the tropics? In the Arctic?

10. Thermoelectric devices have, thus far, been used more widely in the Soviet Union than in the U.S. Why is this?

11. One architect now thinks it is possible to construct a building two miles high and suggests that a "vertical city" could be housed therein. Other than in height, how would this building necessarily differ from conventional skyscrapers?

12. Manned spacecraft are slowly rotated to help maintain temperature equilibrium inside the vehicles. If this were not done, one side would become very hot and the other side very cold. Can you suggest any use that might be made of this situation on long space flights?

13. The Little Siberia Auto Air Conditioning Company announced a car air conditioner that would use the waste heat produced by the car to power its air conditioner. Thus the unit would neither decrease gas mileage nor rob the car of power. What do you think of their idea?

14. Suppose that you were to meet a creature from space who was well adapted to live on a planet on which the gravitational attraction is five times as great as it is on earth. If this space creature's bone strength, blood-vessel strength, etc., were about the same as ours, describe how he might appear. (Would he be very large or very small, or about our size? etc.)

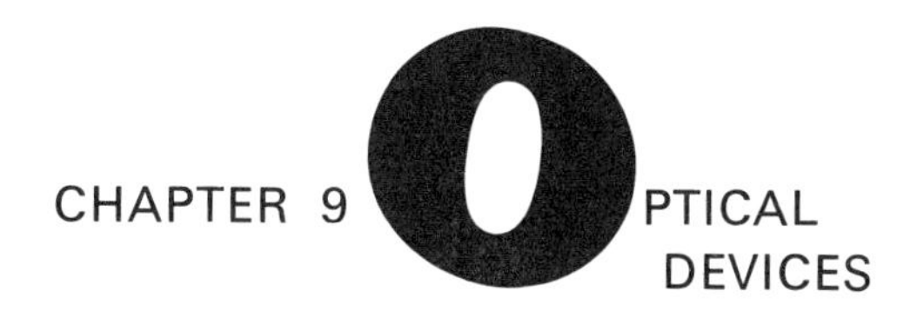

CHAPTER 9 OPTICAL DEVICES

9.1 THE ELECTROMAGNETIC SPECTRUM AND VISIBLE LIGHT

Embedded within the entire electromagnetic spectrum is a narrow band of frequencies which may be detected by the human eye. The highest of those frequencies to which the eye is sensitive is less than twice the frequency of the lowest. Thus the visible spectrum does not span even one whole "electromagnetic octave." Also included within the electromagnetic spectrum are those wavelengths used to carry radio and television broadcasts—these wavelengths are all much longer than light—as well as the very short wavelengths of x-rays and gamma rays. Figure 9.1 shows the relative positions of some of the more familiar portions of the electromagnetic spectrum. (The product of a wavelength and its corresponding frequency equals the speed of light in free space. Frequencies are now expressed in *hertz,* named after the physicist Heinrich Hertz. The hertz replaces the longer term, "cycles per second.")

The visible region lies between wavelengths of 400 and 700 millimicrons (mμ). (One millimicron is 10^{-9} meter.) The shortest wavelengths we can see are those we call violet, while the longest are the ones we know as red. Those wavelengths slightly longer than

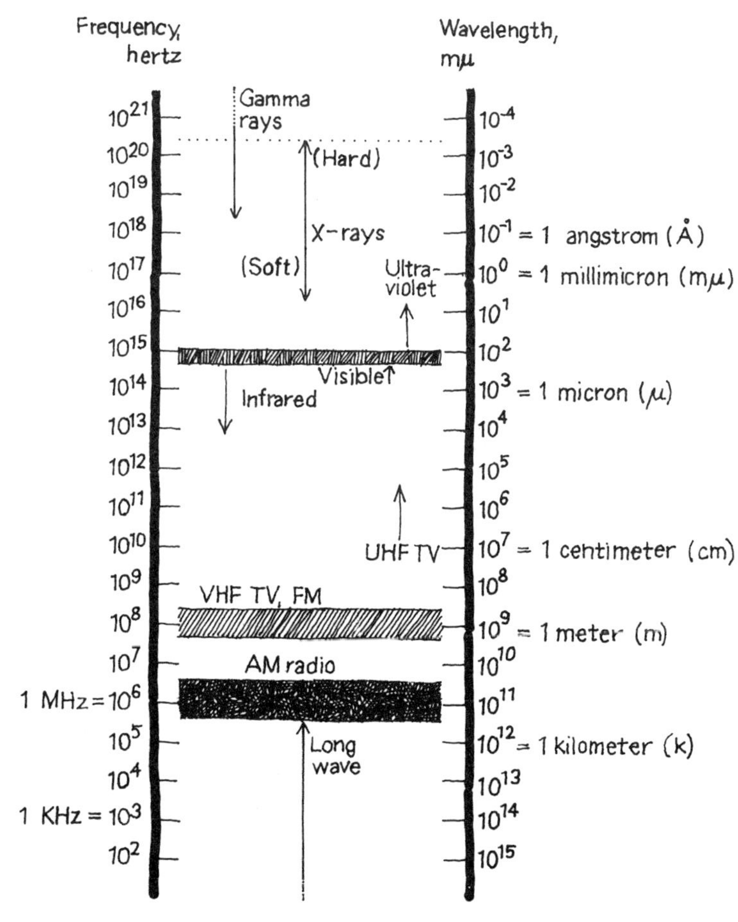

FIG. 9.1

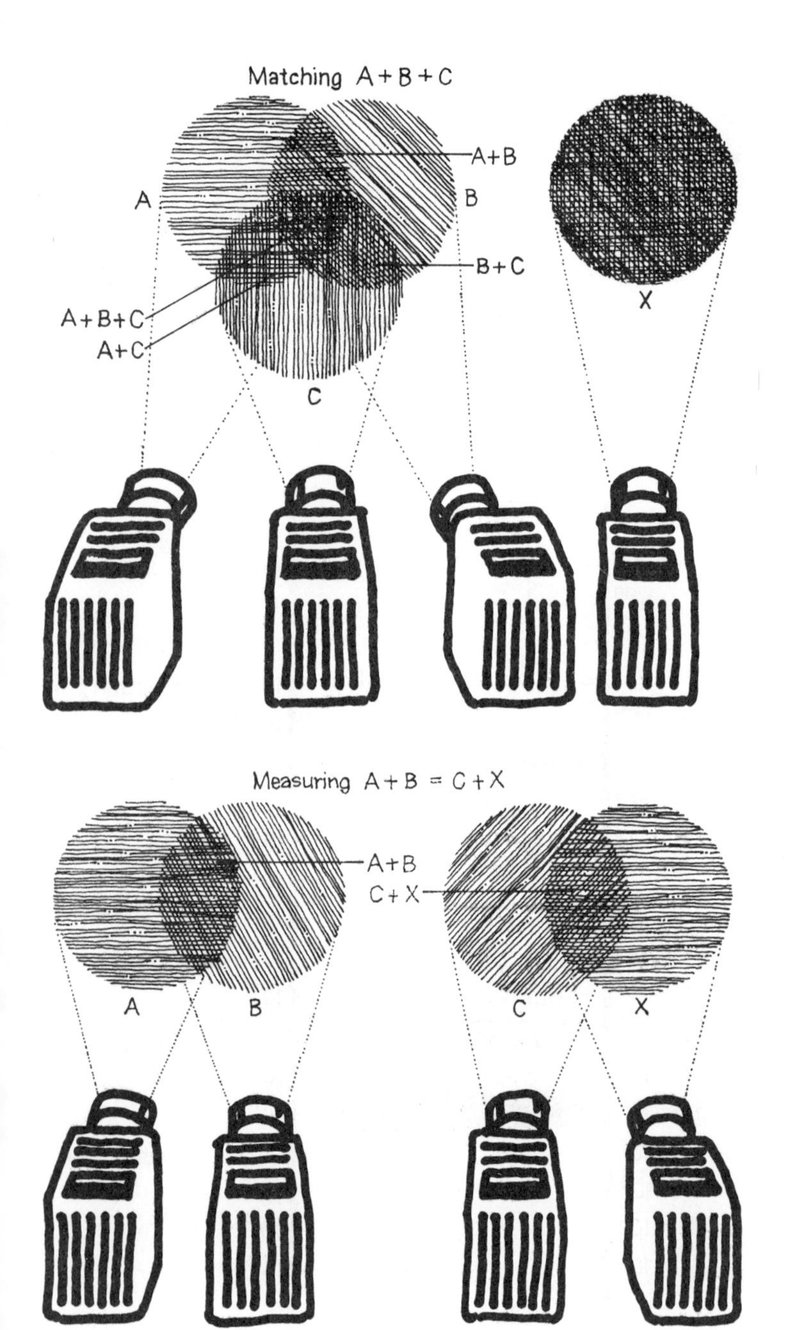

FIG. 9.2

TABLE 9.1

Color	Wavelength range, mμ
Violet	Less than 450
Blue	450–500
Green	500–570
Yellow	570–590
Orange	590–610
Red	Greater than 610

the visible portion of the spectrum make up the *infrared*, while those slightly shorter than visible radiation compose the *ultraviolet*. Table 9.1 lists the approximate wavelengths of the various spectral colors.

Although the colors listed in Table 9.1 may be described in terms of purely physical characteristics (their wavelengths), the sensation of color that is perceived by an observer involves the complex reactions of the eye and the brain to the incoming light waves. The color-observation process differs, in at least one important respect, from the way in which one hears music: if two notes of different frequencies are played simultaneously (on a piano, for example), the listener is aware that there are two distinct sound frequencies present; yet the mixing of two colors produces only a single, third color.

It has often been said that there are three "primary" colors, and any color that may be visualized by the eye may be formed by a mixture of these three primaries (Fig. 9.2). As we shall see shortly, this statement is not quite correct. It is true that a wide range of

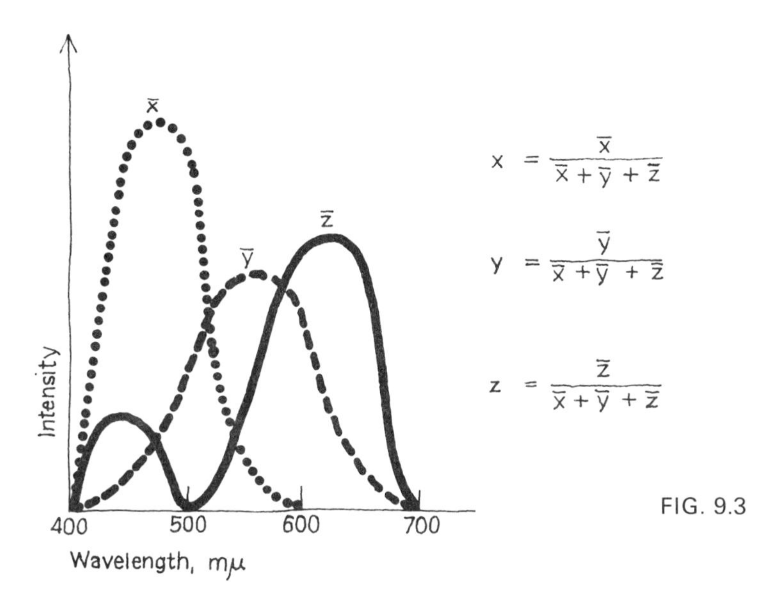

FIG. 9.3

colors may be formed by the mixture of three given colors, but not all colors may be exactly *matched* by any given set of three "primary" colors. Although not all colors may be matched by any three given colors, any color that cannot be matched may be *measured* by a set of three chosen colors. By the measurement of a color which cannot be matched we mean that that color plus one of the three "primary" colors—mixed in some proportions—will be identical to some mixture of the remaining two "primary" colors. Thus any color may be at least specified, or measured, in terms of the three "primary" colors.

For purposes of precisely defining colors, the International Commission on Illumination* has defined three standard colors, which need not be real colors at all. Figure 9.3 shows the intensities of these colors in the various regions of the spectrum. Let us call these curves $\bar{x}$, $\bar{y}$, and $\bar{z}$. Each of these curves is, of course, a function of the wavelength, and could be represented analytically if we wished, but we shall not be concerned with this now. We could measure any color by the proper addition of the colors represented by these curves, but we can present this information somewhat more concisely by defining three other color curves, x, y and z, which are defined in terms of $\bar{x}$, $\bar{y}$ and $\bar{z}$. If we set the condition that $x+y+z = 1$, always, we have four equations and but three unknown quantities. Thus one of these quantities may be eliminated. We choose x and y as the two terms with which we shall deal from here on.

In terms of x and y, all the visible region of the spectrum may be represented as shown in the frontispiece. Note that the pure spectral colors are on the outside of the curve that is shaped like an inverted horseshoe. Somewhere

* This is the English translation of *Commission Internationale des L'Eclairage,* abbreviated C.I.E.

inside this region is that color we know as white. Note that the white point is defined, for we have to pick a point and consider it "pure white." If the point 0 is chosen as the white point, we may then define the *purity* of the various colors in terms of the distance from this point to the locus of the curve.

For example, a wavelength of 540 mμ represents a rather nice spectral green, Now the color represented at point *A* on the diagram will also appear green, and, since it lies on the curve of constant hue between the white point and 540 mμ, it will appear green, peaked at 540 mμ. But it will not be as "green" a green as the color represented by the point on the outside of the curve. Thus the purity of this color is defined as the percentage of the distance that color lies between the white point and the loci of the pure spectral colors. The purity of a spectral color is 100% and the purity of white is, of course, zero.

On this same diagram we may illustrate the result of mixing any two colors. If colors *D* and *E* are mixed, they will produce a color lying on the straight line connecting *D* and *E*. The position of this point on the line will depend on how much of each color is used. Identical amounts of *D* and *E* will produce a color represented by the point halfway between *D* and *E*.

Note that while most colors may be formed by mixing white and a pure spectral color, those in the shaded part of the enclosure cannot be formed by mixing white and a pure spectral color. These colors, known as *nonspectral* colors, contain the purples.

Now it should be obvious why we cannot match any color by mixing a single set of three "primary" colors. The widest range of colors may be matched by using colors at the vertices of the triangle shown in the frontispiece. These colors correspond, as you might expect, to a red, a green, and a blue. But those colors lying outside this triangle cannot be made by mixing these three colors. A color such as *N* may be formed with the same dominant wavelength as the color *M,* which is a pure spectral color, but the color purity of *N* is only about 50%. The triangle shown cannot be formed from points outside the curve shown, because the region shown contains all the visible colors.

The preceding discussion has to do with the formation of colors by adding them together. The mixing of colors that we have described might be accomplished by transmitting light through transparent colored filters. But it is important to realize that most of our sensation of color depends on a *subtraction* process. If white light illuminates a piece of paper and the sheet of paper appears to be white, this is an indication that the paper is reflecting rather uniformly in all parts of the visible spectrum (Fig. 9.4). If the paper appears red, what has happened is that the blue and green wavelengths have been absorbed, leaving only a strong reflection of the red wavelengths.

The apparent color of any object viewed by reflected light depends on which wavelengths are absorbed and which are reflected by that object. It must be emphasized here that in order to perceive the "true" color of any object, one must view it by light that contains representatives of all wavelengths. Most ordinary incandescent lighting (for example, the common light bulb) is deficient in the blue part of the spectrum, while mercury vapor lamps (widely used in outdoor lighting) are deficient

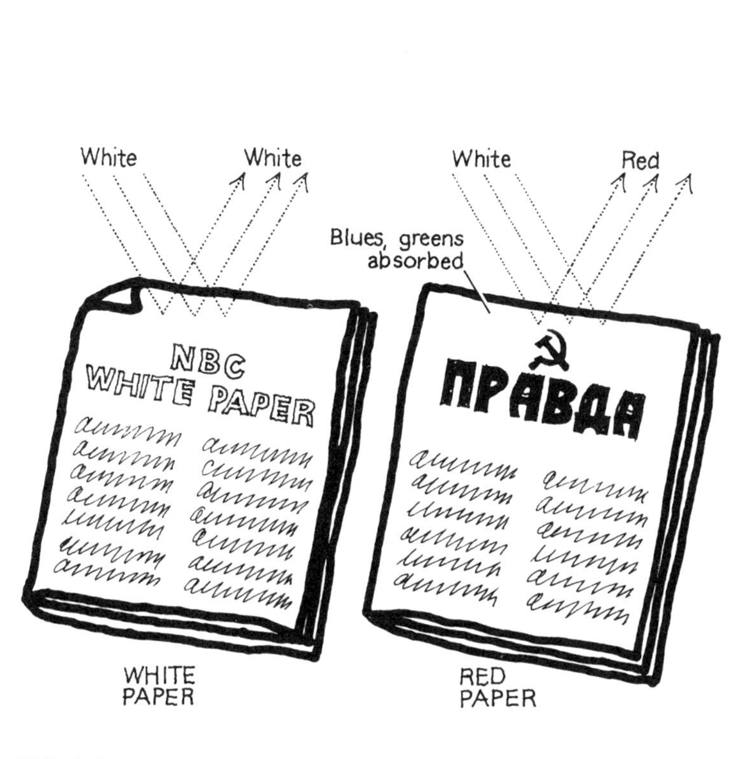

FIG. 9.4

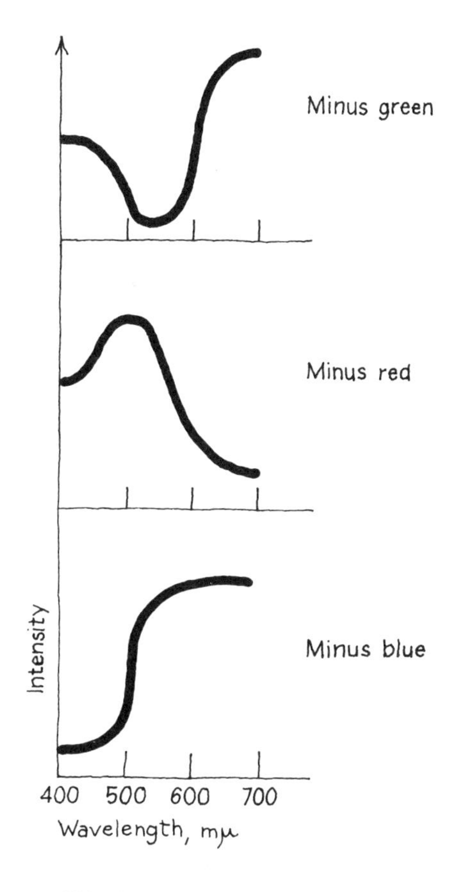

FIG. 9.5

in the red part of the spectrum. Objects do not appear to be the same color when viewed in these types of lighting as they do when illuminated by a pure white light.

We may quantify this discussion of color perception by reflected light by defining three subtractive color primaries which serve the same sort of purpose as the additive primaries. Figure 9.5 shows the wavelength/intensity plots for minus green (magenta), minus red (cyan), and minus blue, which appears yellow. If we wish to print the color that appears red, we print (minus green) plus (minus blue), which is red. Green is obtained by printing (minus blue) plus (minus red). Black, which is the visual effect observed when *all* wavelengths are absorbed, cannot normally be satisfactorily represented by the addition of all three subtractive primaries. It is usually printed separately, leading to the "four-color" process commonly used in color printing.

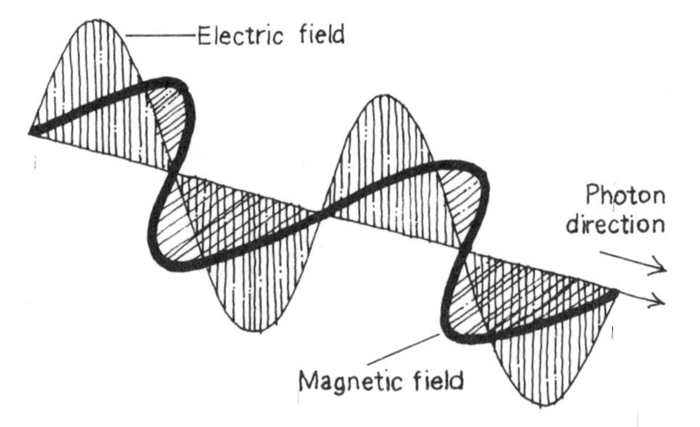

FIG. 9.6

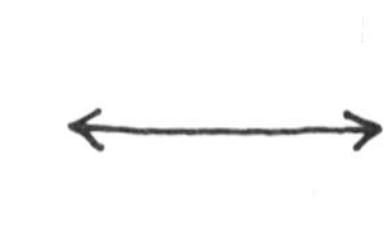

FIG. 9.7 Head-on view of electric field vectors.

9.2 POLARIZATION

We have been using the term "electromagnetic radiation" throughout this chapter. Now let us point out *why* this term is used. Electromagnetic radiation is composed of two fields: an electric field and a magnetic field (Fig. 9.6). These fields may be represented at any given point, for a single photon, by two vectors: an electric vector and a magnetic vector. These vectors are always at right angles to one another, and vary sinusoidally. So, to describe the propagation of the electromagnetic wave, we usually need discuss only one of these vectors.

Ordinary sources of electromagnetic radiation emit photons that are of random phases and have their electric and magnetic vectors aligned in all directions. Figure 9.7 shows the distribution of electric vectors as seen head-on from such a light source. When these vectors are aligned so that they are all in a given plane, we say that the light has been *plane polarized*. The plane in which all the electric vectors lie is known as the *plane of polarization*.

Certain materials have optical properties that make it possible for them to transmit only light that is polarized in a given plane. If a beam of unpolarized light is passed through a filter that is polarized vertically, only the vertical components of the electric vector can pass through this filter. This means that the intensity of the light transmitted from a beam that is unpolarized will be cut in half. If this (now) polarized beam of light is then incident on a filter that is horizontally polarized, no light at all will be transmitted (provided that both filters are 100% effective).

When ordinary unpolarized light is reflected from a surface, the reflected light is partially polarized (Fig. 9.8). This effect reaches a maximum for each reflecting material at an angle that is known as the *polarizing angle*. This reflected light has its electric vector polarized horizontally. Thus, when viewed through material that is vertically polarized, most of the reflected light is not transmitted. This phenomenon is responsible for the effec-

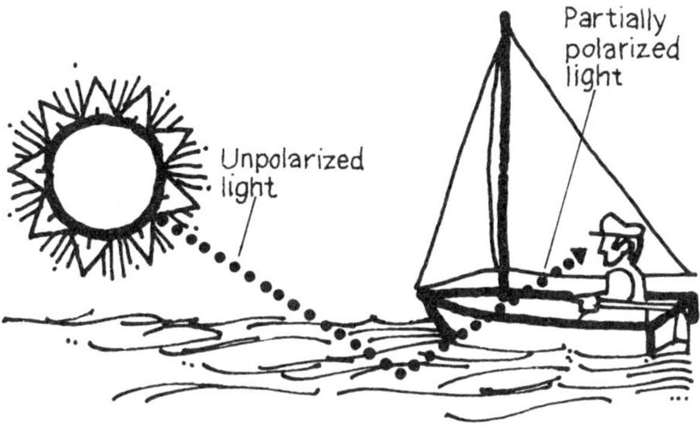

FIG. 9.8

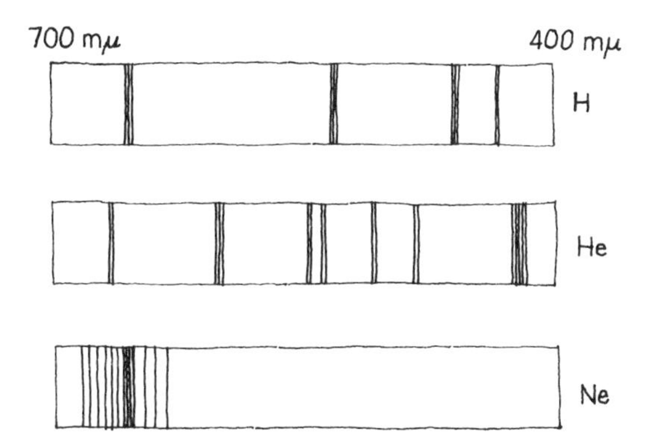

FIG. 9.9 Visible gas emission spectra.

tive filtering of glare reflected from highway and water surfaces by the polarized material used in sunglasses.

The sky appears blue in the daytime because the white sunlight is scattered—i.e., absorbed and re-emitted—by molecules in the atmosphere. This scattering depends on the reciprocal of the fourth power of the wavelength of the light involved. Thus wavelengths in the blue part of the spectrum are scattered much more efficiently than those in the red. In the scattering process, the re-radiated light is polarized. If a polarizing filter is placed in front of the lens of a camera taking color pictures, the sky will be made to appear much darker in the pictures, without upsetting the color balance of the scene.

9.3 LASERS

Most sources of light, as we have noted, emit radiation that is of random phase, or incoherent light. Light that is emitted in phase—coherent light, such as is produced by lasers—may be used to produce a number of important and useful effects that cannot be duplicated by ordinary incoherent light.

The word *laser* is an acronym, a word made up of the first letters of its descriptive phase: *L*ight *A*mplification by *S*timulated *E*mission of *R*adiation. The laser is the second generation of such devices, for it was constructed only after similar devices, known as *masers*, had been used in the microwave regions. For some time after their introduction, lasers were known as "optical masers," but the laser is now so important that the name laser is used even for devices that operate in infrared or ultraviolet parts of the spectrum.

Operating Principle of the Laser

To fully appreciate what is involved in the operation of the laser, it is necessary to briefly discuss the emission of incoherent light. If a gas (Fig. 9.9) is placed in a glass tube and a high electrical potential applied across it, the

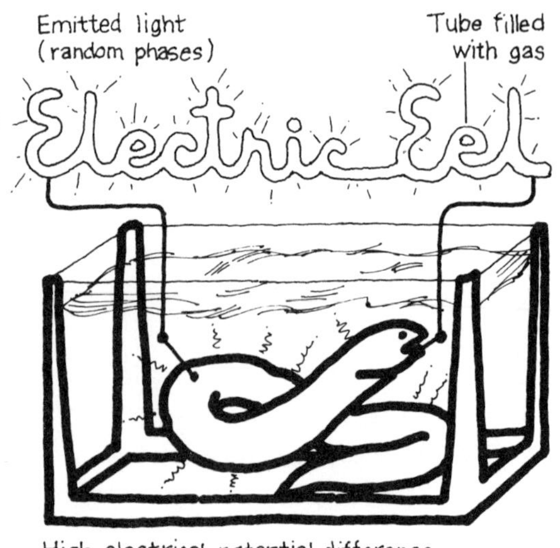

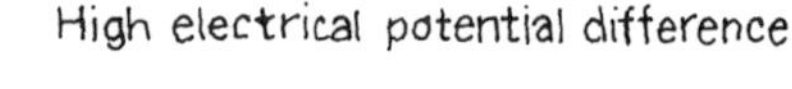

FIG. 9.10

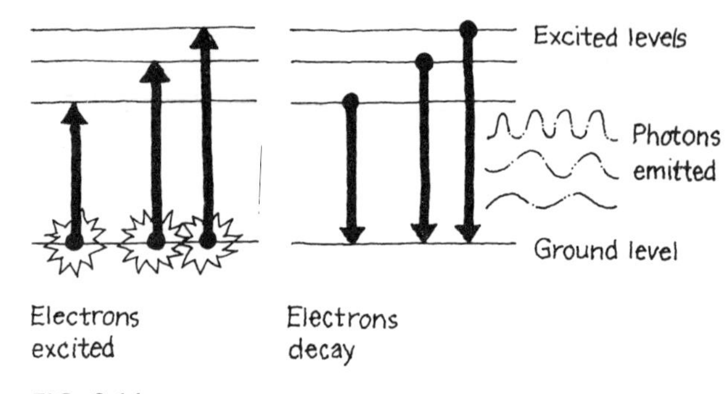

FIG. 9.11

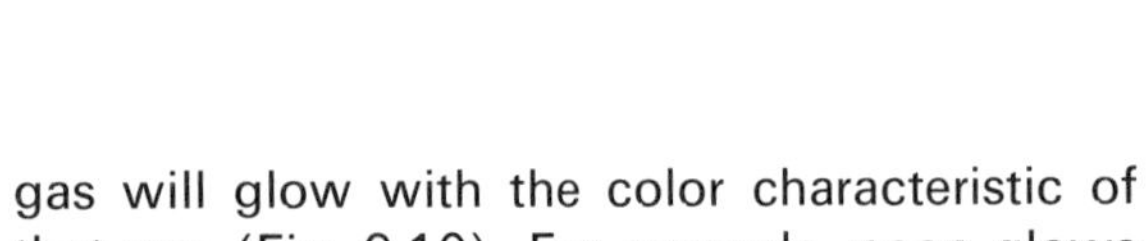

gas will glow with the color characteristic of that gas (Fig. 9.10). For example, neon glows with the familiar red we see in neon lights, while other gases may be used to produce other colors.

If this emitted light from any single element were examined in a spectroscope (an instrument which separates the various wavelengths of light) we would see that the light spectrum emitted by the gas is not continuous, but consists of discrete spectral lines. Neon gives its characteristic red glow because almost all the lines are in the red part of the spectrum, while oxygen has lines primarily in the blue. The identification of these lines is an important function of spectroscopy. The element helium was first discovered from spectral lines emitted from the sun (hence its name: Greek *helios* = sun) before the gas was isolated on earth. Law-enforcement agencies have used the spectral lines emitted from chips of automobile paint to identify the make and year of manufacture of vehicles involved in hit-and-run accidents.

The emission of the different wavelengths of light that produce the spectral lines of an excited gas comes about in the following way. The electrons in the gas are each normally in the lowest energy level available to them in orbit in the atoms of the gas. When the gas is excited by passing an electric current through it, electrons are raised to higher energy levels (Fig. 9.11). They quickly decay back to their lower, ground levels, emitting photons of energy precisely equal to the energy difference between the excited level and the ground level. Since only certain excited energy levels are available to these electrons, the photons emitted have discrete wavelengths, which may be observed in the spectrum of the light emitted.

The process in which this takes place involves electrons starting from their ground states, being excited to higher levels, and spontaneously decaying back to their ground states. Since the decay to the ground state is spontaneous, the phases of the emitted photons are all random.

In a laser, the emission process is one of *stimulated emission*, which makes it possible to have photons emitted that are coherent, or in phase. The secret of constructing a laser is to find a way of keeping electrons in excited levels long enough to have their photon emission stimulated, or triggered, rather than happening spontaneously. This is possible if there exists an arrangement of energy levels such as those shown in Fig. 9.12.

Electrons are raised to excited states by a pumping process. This may be done by illuminating the laser with an external light source of the proper frequency, in which case the process is known as *optical pumping*, or raising it by an electrical discharge, which causes *electrical pumping*. The level *A*, to which electrons are pumped, is a short-lived state and the electrons will not remain there long, but spontaneously return to the lower energy levels. Some of these electrons will go from *A* directly back to *C*, but not all of them will return to the ground level quickly, as there is a level such as *B*, which may hold an electron in a *metastable state*. A metastable state is one

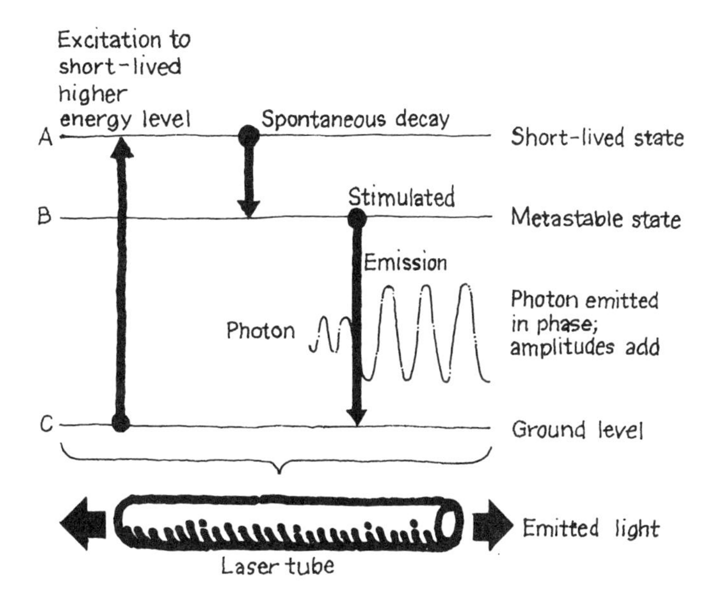

FIG. 9.12

for which quantum-mechanical selection rules make it difficult for an electron to decay directly to the ground state. To return to the ground state, this electron would normally have to be excited to a higher energy level, from which it could return to the ground state. Thus a metastable state is a relatively long-lived one.

Although an electron in the metastable state does not normally return to the ground state spontaneously, it may be stimulated to emit a photon having an energy exactly equal to the difference in this energy level and the ground state, and thereby return to the ground state. This emission is stimulated by another photon whose frequency precisely corresponds to the one that is emitted. When emission is stimulated in this way, the emitted photon is exactly in phase with the photon that stimulated the emission. Thus the amplitudes of the two photons add constructively.

A passing photon of energy exactly equal to the difference between the ground state and the metastable state might equally well excite a ground-state electron into the metastable state, with the photon being absorbed in the process. The probabilities of these two competing processes (stimulating the in-phase emission of another photon or the absorption of the first photon) depend on the electron populations of the ground state and the metastable state. If there are more electrons in the ground state than in the metastable state, then more photons are absorbed than emitted, and the emission process is damped out. But if there are more electrons in the metastable level than in the ground level, there is a buildup of emitted radiation, and the laser will "fire."

Normally there are always more electrons in the ground level than in any excited level. Thus the condition which is necessary for the operation of a laser is that there be a *population inversion*: The energy level above the ground state must contain more electrons than the ground state.

In a gas laser, the laser gas may be in a tube, pumped optically by a surrounding light source. As the stimulated emission process takes place, more and more electrons in the metastable level are caused to return to the ground state by the stimulated emission of radiation. The (stimulated) emitted radiation builds up with the emission of each (in-phase) photon. The ends of the laser are coated with a special reflective material so that most of the light is reflected back into the laser, with only a small portion of the radiant energy released externally.

Lasers may utilize gases, solids, or even liquid organic materials. A single laser may be constructed that can be modified to emit light of any one of several available frequencies. Lasers may operate continuously, though the most powerful ones operate in pulses.

Uses of Laser Devices

We have noted that the electromagnetic radiation from a laser is coherent: that is, all the photons are in phase. Because of this, the light output can be focused onto a much smaller spot than light from an incoherent source. In the case of ordinary light, it is quite impossible to focus a beam into a space any shorter than a single wavelength of the light. For a laser this limitation is removed, and the area into which the beam may be focused is limited only by the optics involved. This means that the *energy density* of the output beam from a laser may be many times greater than that of the most powerful conventional source, and lasers may be used for cutting and drilling holes in materials such as steel.

In the early days of laser work experimenters sometimes gave lasers tongue-in-cheek power ratings in "gillettes"—indicating the number of steel razor blades the laser beam could cut through. Typical lasers with power ratings in only tens of watts may be used to cut and drill in very hard materials, because the energy of the beam is concentrated into extremely small areas.

It is interesting to theorize about future high-powered lasers that may be used to dig tunnels, cut and weld thick steel plates, or even destroy incoming ICBM's. A new genera-

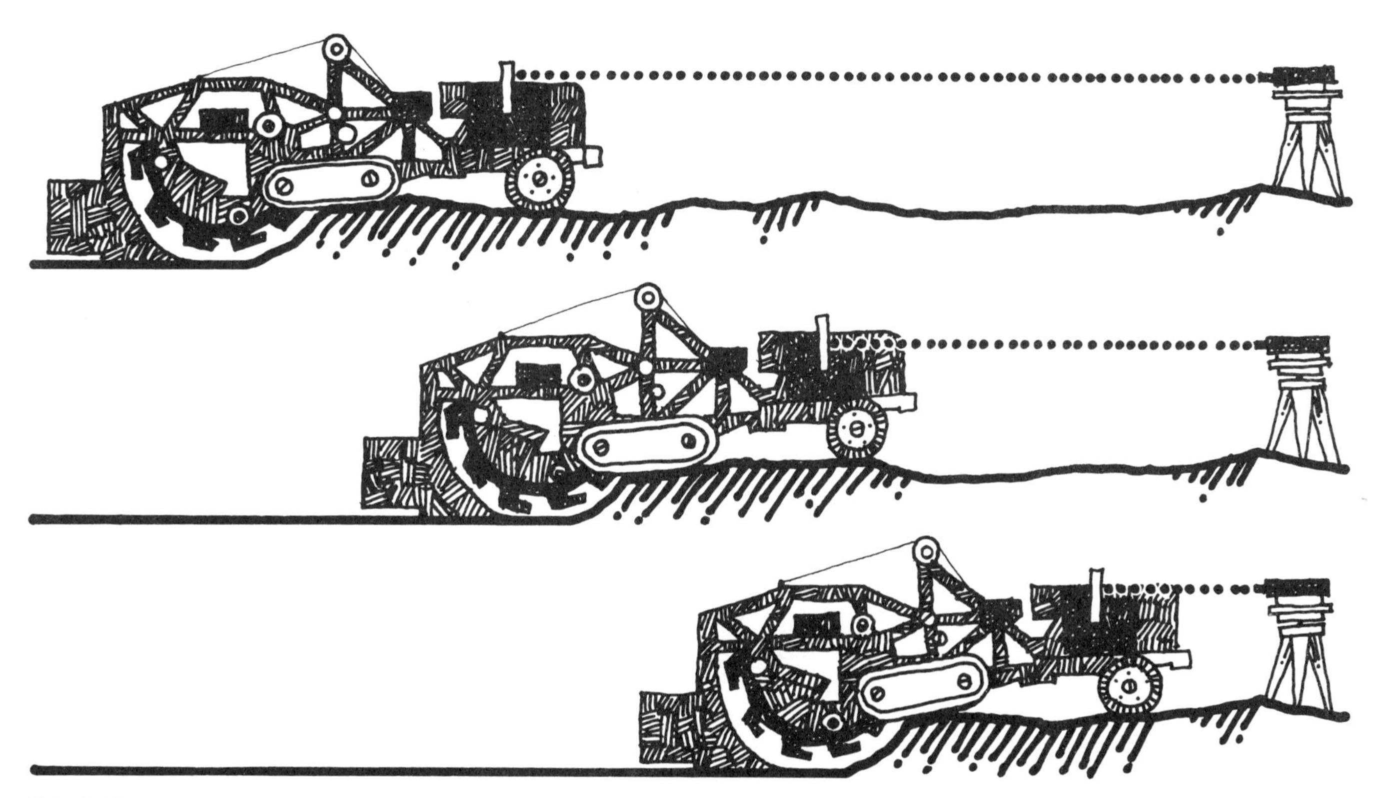

FIG. 9.13 Laser alignment device for automatic control of depth of an excavation.

tion of lasers that use *thermal pumping*, driven by rocket engines, may be able to accomplish these and other objectives. Continuous power outputs as great as 60 kilowatts have already been announced, and still greater power outputs are available in pulsed operation. Lower-powered lasers are now being used as cutting devices in applications such as the construction of miniature solid-state electronic components. Machining devices that use laser beams for drilling, welding, cutting, trimming, and other fabrication work are now available (Fig. 9.13). Another technique that has been demonstrated, though not commercially developed, is the use of a laser as a typewriter eraser. The beam is aimed at the incorrect symbol and, as energy is absorbed from the beam, the incorrect symbol is vaporized. The carbon symbol absorbs heat from the laser beam much more quickly than the paper, which is unaffected during the brief time that the beam is turned on.

Laser applications in medicine include using laser beams to "weld" detached retinas. Lasers have also been used experimentally to remove

Two laser beams, one originating at Kitt Peak National Observatory (near Tucson, Arizona) and the other at Table Mountain Observatory (near Los Angeles), as seen by television from Surveyor VII (January 1968) on the moon. (Photograph courtesy of NASA)

tattoos and other skin blemishes. Future techniques may involve the use of laser beams as cutting devices in surgery. The laser beam would not only have the advantage of making possible micrometer-precise control in making an incision, but the beam would cauterize the area surrounding the cut. Since nothing but photons would touch the flesh, laser incisions would involve the most sterile surgical procedure imaginable. The development of dental techniques in which lasers would be used for drilling and cutting teeth is anxiously awaited by many squeamish patients.

The property of spatial coherence makes it possible for a laser beam to remain very compact and well defined, even after it has traveled great distances. Light from ordinary extended light sources spreads out very rapidly after it leaves its source, even when focused into a spotlight type of beam. Because the beam of a laser remains well defined, the energy density of the beam remains high. In January 1968, two laser beams from earth were photographed by Surveyor VII from the moon (see photograph). Despite the fact that each of these beams had a power of only about *one* watt, from the moon they appeared as the brightest objects on the darkened earth. Laser beams sent from earth have also been bounced off a reflector on the surface of the moon and observed again on earth.

The narrow definition and small spreading of laser beams makes them useful for alignment procedures. Already lasers have been used for tasks such as aligning the 3-kilometer Stanford Linear Accelerator in California to an accuracy of 0.5 mm and aligning the jigs used in constructing the 38-m-long wings of the Boeing 747 jumbo jetliner. A laser beam served to guide dredging operations in San Francisco Bay during construction of the Bay Area Rapid Transit System. Countless other uses of laser beams in surveying and ranging operations are now under way. Possibly the most scientifically rewarding of these uses involves experiments now being conducted in which light is bounced from an earth-based laser off the retrodirective reflector left on the surface of the moon by the Apollo 11 mission. These measurements pinpoint variations in the earth–moon distance and may make possible the measurement of any variation of the strength of the gravitational interaction itself. Precise answers to questions asked about continental drift—the motion of the earth's land masses relative to one another—may also be obtained from this type of measurement.

Because lasers operate in the optical region, their use frequencies are about 10^{14} Hz. The fact that these are extremely high frequencies is important because of the information-carrying capacity of such high frequencies. We express information-carrying capacity in terms of binary "bits" of information relayed. *Binary notation* makes use of only two digits: 0 and 1.

TABLE 9.2

Decimal	Binary
0	0
1	1
2	10
3	11
4	100
5	101
6	110
7	111
8	1000
9	1001
10	1010
11	1011
12	1100
13	1101
14	1110
15	1111
16	10000

$$\begin{array}{rll} 13 = & 10 = 1\times10^1 \\ & \underline{+3} = 3\times10^0 \\ & 13 \end{array}$$

$$\begin{array}{rlll} 13 = & 1000 = 1\times2^3 & = & (8) \\ & +100 = 1\times2^2 & = & (4) \\ & +\ \ 00 = 0\times2^1 & = & (0) \\ & \underline{+\ \ \ \ 1} = 1\times2^0 & = & \underline{(1)} \\ & 1101 & & (13) \end{array}$$

This property makes the binary number system particularly adaptable to electronic encoding and storage devices, since one of these digits may represent "off" and other "on," or they may represent the positive and negative slopes of electronic signal pulses. For your convenience, Table 9.2 shows the first few numbers in both decimal and binary systems. If you are not already familiar with the binary number system, note that this system uses base 2, just as base 10 is used in our ordinary decimal system.

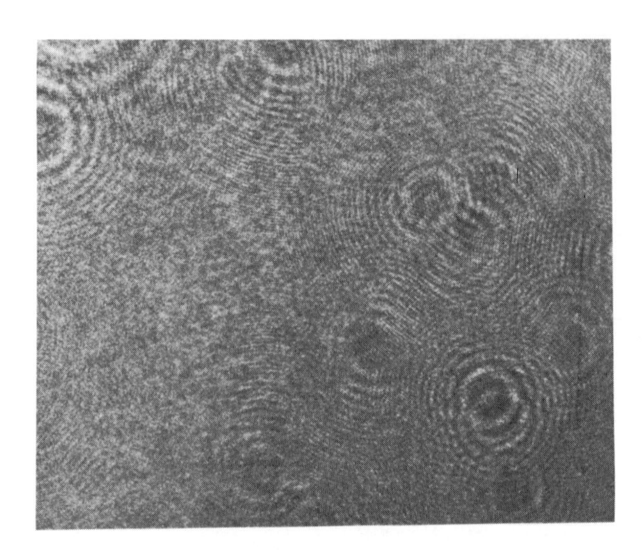

A hologram recorded on photographic film illuminated by ordinary light. You can see that the hologram itself bears no resemblance to the original scene. It records, but it contains enough information to allow the reconstruction of a real three-dimensional image of that scene.

To represent any one of the numbers between 0 and 9, you may see that you will need 4 bits of information. Of course those 4 bits actually enable you to represent 16 numbers and—in general—n bits represent 2^n numbers. When one is transmitting information rapidly, the number of bits required per second depends on the complexity of the information being transmitted. Speech communication requires the transmission of about 7×10^4 bits/second, while black-and-white TV requires about 4×10^7 bits/second. The higher the frequency of the carrier used, the greater the number of bits/second that may be transmitted. Thus higher frequencies mean greater information-carrying capabilities.

Long-distance telephone signals are now carried from point to point over microwave relay networks in most areas of the U.S. If these signals were carried in the optical region by lasers, we would immediately have available 100,000 times our present signal-carrying capacity.

Laser signals are not now being used for communications purposes. One problem that must be completely solved before this is possible is the rapid and efficient modulation of laser signals. Another problem involves the fact that smog or fog can greatly attenuate a laser signal. Either transmission of laser signals in the infrared part of the spectrum or confinement of laser transmission to closed "light-pipes" could eliminate this difficulty. In any event, laser communications hold very great promise for communications applications.

Perhaps the most dramatic laser application at present is in holography. A *hologram* is not simply an ordinary photograph, exposed by light from a laser, for it is possible for the viewer of a hologram to see a real three-dimensional image of the object depicted in the hologram.

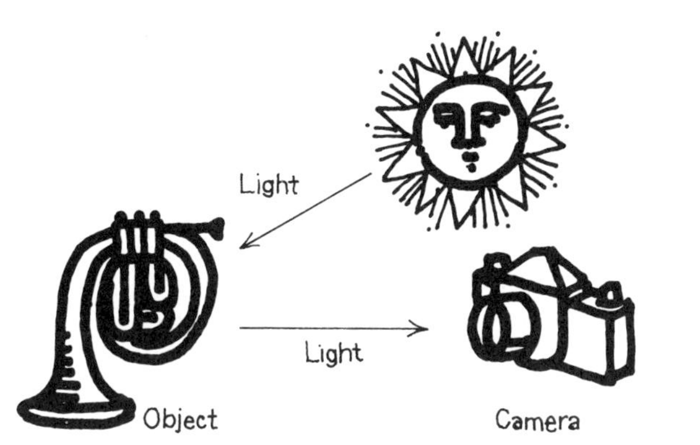

FIG. 9.14 Ordinary photography.

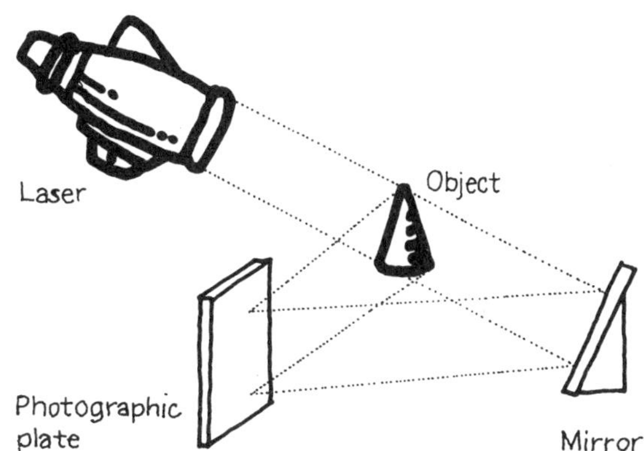

FIG. 9.15 Making a hologram.

To illustrate the difference between holography and ordinary photography, let's first review the process by which an ordinary photograph is made (Fig. 9.14). In photography, an object is illuminated by normal, incoherent light. A part of this reflected light is focused onto a film and its image recorded on the film. When examining a photograph the observer may view only a two-dimensional representation of a three-dimensional scene, and his angle of viewing is forever fixed at whatever angle the camera was positioned when the photograph was exposed.

In order to expose an ordinary photograph, light from the object being photographed must travel from the object, through the camera lens, and be focused on the film. After the photons used in a particular exposure have left the object being pictured, the object may undergo some change—for example, a human subject may move—though that change no longer has any effect on the picture that is recorded, once the group of photons that expose the picture are on their way to the camera. Thus, anyone who later views this picture will see the object photographed only as it was represented by the particular group of photons that entered the camera lens and exposed the film. If the viewer had some means of observing a group of photons just like the ones that originally came from the object, it would be possible for him to see the same thing he would see if he were looking at the original object at the instant the photograph was made. This is precisely what is possible in holography.

To record a hologram, one illuminates an object with the coherent light from a laser beam (Fig. 9.15). Both the light reflected from the object and the light coming directly from the laser are made to arrive together at a photographic film. As the two beams of

A photograph of an image reconstructed from a hologram of a model train. The image was produced by illuminating the hologram with a laser beam. The reconstructed image was three-dimensional and contained more information about the original scene than this photograph of that image.

light come together at the film, there is both constructive and destructive interference by these light waves, with their resulting interference pattern recorded on the film. This is what makes up a hologram: an interference pattern of light waves recorded in a photographic emulsion. Note that no lenses need be used in making a hologram, and there is no way for it to be "out of focus" in the way that an ordinary photo might be.

To view the hologram, the viewer illuminates it by light from the same sort of laser used in making the exposure (Fig. 9.16). He then sees a reconstruction of the wavefronts that originally exposed the hologram. That is, he sees light waves coming toward him that appear just as they would if he were viewing the actual object instead of a hologram. This means that he can look at the object from different angles and gain perspective of it. If part of the scene

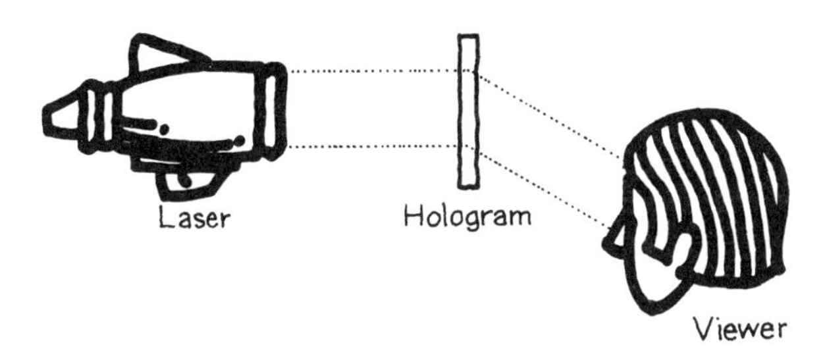

FIG. 9.16 Viewing the hologram.

is blocked by something else, the viewer, by moving his head to one side, simply looks around the part of the representation that is in his way. (Have you ever been in a group picture, only to find later that someone's head had been between you and the camera? If you viewed a hologram of the same scene, you could see your image by simply looking around the person who was in front of you.)

There are other important differences between holograms and photographs. If you tear off a part of a photographic negative, you will have left only part of the picture from which it came. But each small part of a hologram contains the entire picture. Granted, the quality of the rendering of the scene is diminished when one views a small part of a hologram, but the entire scene is still there. This means that a hologram will not be ruined when a portion of it is spoiled or destroyed.

Another interesting characteristic of holograms is that they form their own negatives. If a hologram is copied onto negative film, the copied hologram produces the same picture as the original without reversing white and black. This is due to the fact that only the reconstructed wavefronts are involved, and the difference between an image and its negative is not meaningful in this application.

Finally, a single photographic plate may contain a large number of holograms. To store more than one hologram on a plate, one has only to tip or tilt it slightly before the next exposure. A hologram "camera" could record several holograms on the same sheet of film. This property suggests the potential use of holography in information storage and retrieval applications.

At any given time, a laser operates with a light output of a single frequency. Therefore a hologram made and viewed by a single laser is necessarily monochromatic. However, holograms may be formed and viewed with accurate color rendition by using more than one laser beam, so long as the beams represent different parts of the color spectrum.

FOR MORE INFORMATION

A. C. S. van Heel, *What is Light?*, translated by J. L. S. Rosenfeld, New York: McGraw-Hill, 1968. (In paperback.)

S. Tolansky, *Revolution in Optics*, Baltimore, Md.: Penguin, 1968.

Emmett N. Leith and Juris Upatnieks, "Photography by Laser," *Scientific American*, June 1965, page 24.

Stewart E. Miller, "Communication by Laser," *Scientific American*, January 1966, page 19.

J. A. Giordmaine, "Nonlinear Optics," *Physics Today*, January 1969, page 39.

Michael W. Berns and Donald E. Rounds, "Cell Surgery by Laser," *Scientific American*, February 1970, page 99.

QUESTIONS

1. What is the range of frequencies to which the human eye is sensitive?
2. Which has the longer wavelength, red or blue light?
3. Do TV stations broadcast at wavelengths shorter or longer than those detectable by the eye?
4. Can a set of primary colors always be formed to match any given color?
5. What do we mean when we say that three primary colors may be used to measure another color?
6. What is meant, physically, by color purity?
7. What is a nonspectral color? Give an example.
8. Illustrate why no three primary colors can be used to match any given color.
9. Discuss the differences between additive and subtractive color processes.
10. What does (minus green) plus (minus blue) produce?
11. Why are comic strips printed in a four-color process? What is the fourth color?
12. What do we mean by plane-polarized light?
13. How can one detect polarized light?
14. What does the reflection of light have to do with its polarization?
15. What has the scattering of light to do with blue skies and red sunsets?
16. What color would the sky appear if there were no atmosphere?
17. What does the word *laser* stand for?
18. What is the difference between stimulated emission and normal emission of photons?
19. What is optical pumping?
20. Why is a population inversion needed in a laser?
21. What is a five-gillette laser?
22. List some highly desirable medical applications of lasers.
23. List some practical applications of lasers that have already been made.
24. Describe some of the advantages and disadvantages of a laser communications system.
25. How is a hologram different from an ordinary photograph?
26. List some possible uses of holography.
27. Can holograms be produced in color? How?
28. Why can a laser beam be reflected from a mirror on the moon and detected again on earth when a spotlight that puts out much more radiant energy than the laser cannot be seen?

PROBLEMS

1. An "invisible paint"—one which would render anything painted with it invisible—has been the object of much speculation. What are the optical properties such a paint must have? Can this paint ever be developed?

2. Meat counters in groceries are normally lighted by lamps that produce strong reds and are deficient in blues. Mercury vapor lamps make reds look like purple, but grass look greener. Discuss the potential uses and *abuses* of nonwhite lighting in merchandising, etc.

3. Laser "death rays" may or may not be available for military use one day. What are the operational criteria for such devices? What advantages might they have over conventional weapons?

4. As chief executive officer of Living Color TV, Inc., you must decide whether or not to develop color holographic television sets. *Can* this be done? How? What factors would weigh most heavily in your evaluation of the practicability of making this product?

5. State whatever job or profession you expect to be in five years from now, and describe one possible use of holography in that job. (You may envision holograms that are as large as you wish, in full color, etc.)

6. You are given a number of optical filters, some of which are polaroids and some of which aren't. How can you tell which of these are polaroid materials? How can you determine the polarizing axis for each filter?

7. Discuss the advantages and disadvantages peculiar to laser communications systems for use (a) between nearby points on earth, (b) between distant earth points, with the aid of an earth satellite, and (c) for communications between vehicles in space.

CHAPTER 10 THERMAL PHENOMENA

10.1 TEMPERATURE SCALES

Most persons in the English-speaking world are used to measuring and expressing temperatures in terms of the Fahrenheit temperature scale. This unfortunately awkward scale, which divides the temperature range between the equilibrium point of an ice/water mixture (32°F) and the steam point (212°F) into 180 equal degrees, may seem rather in keeping with systems of units that include inches and feet and yards and miles. However, it happened to be set up as it is because the man for whom the scale was named, Gabriel Daniel Fahrenheit, a German instrument-maker, in 1724 could reach a temperature no lower than zero degrees (F) with a mixture of salt, water, and ice. This temperature became the starting point of Fahrenheit's scale, and most English-speaking people have continued to use Fahrenheit degrees for everyday measurements.

The first person recognized for dividing the temperature scale between the ice and steam points into 100 degrees was a Swede, Andreas Celsius, who, in 1742, suggested taking the steam point as 0° and the ice point as 100°. Fortunately his upside-down temperature scale was not popular, but the idea of 100 intervals between the ice and steam points was adopted with the ice point at 0°. This scale, for years referred to as the *centigrade* scale, is now known as the *Celsius* scale, although we continue to use the abbreviation °C.

Neither of the above temperature scales is very satisfactory for scientific usage, since temperature values used in calculations must be based on the absolute zero of temperature as the starting point. It was Lord Kelvin who suggested that the absolute zero be 0 degrees, and, using the Celsius (née centigrade) degree, the *Kelvin* temperature scale (°K) is named in his honor. For certain engineering usages in the English system of units, however, it is convenient to have a scale that starts at zero, but uses the Fahrenheit degree. This scale is known as the *Rankine* scale (°R). Figure 10.1 shows the relationship between these four systems of temperature units.

For very precise measurements, it is not convenient to use the ice and steam points as reference points in setting up temperature scales, since this would mean that the absolute zero of temperature would have to be measured. The absolute zero is about 273 degrees below the ice point, and at very low temperatures a small error in temperature measurement would lead to a very large percentage of error. The

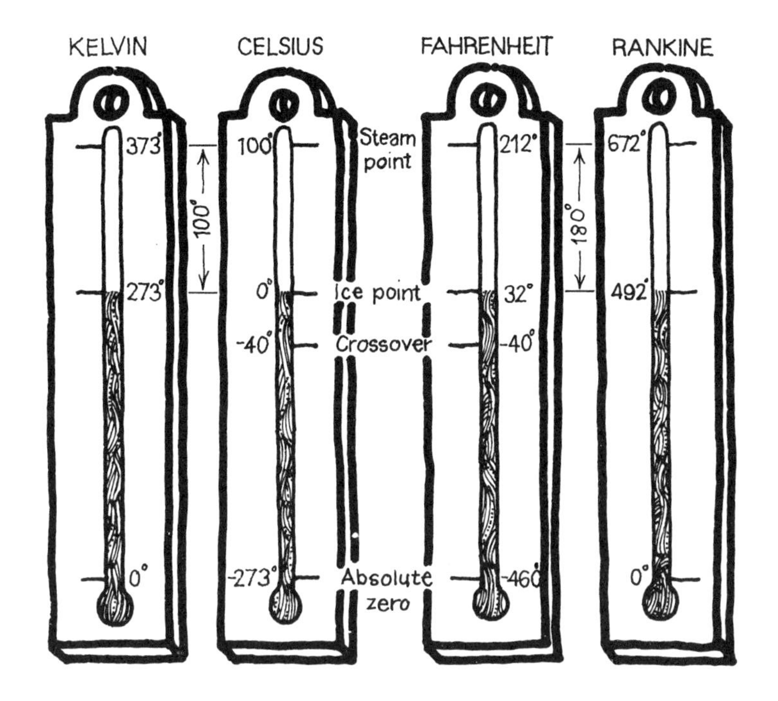

FIG. 10.1 Comparison of temperature scales.

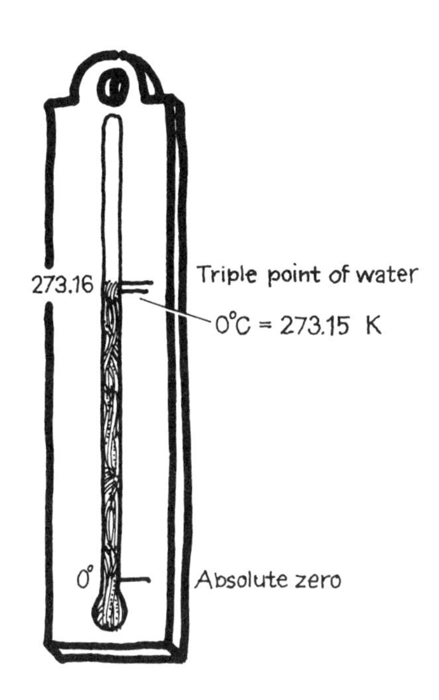

FIG. 10.2 Kelvin temperature scale.

standard now used simply sets the absolute zero of temperature at 0 kelvins (we no longer say "degrees K," but simply so many "K") and the *triple point* of water at 273.16 K. The triple point of water is that point at which water, ice, and water vapor all exist in equilibrium (Fig. 10.2).

10.2 HEAT, ENERGY, AND THERMODYNAMICS

From temperature scales that allow us to deal quantitatively with the concepts of "how hot" or "how cold" objects may be, we now turn our attention to problems associated with changing the temperature of an object. An examination of some common physical phenomena makes it rather apparent that there is a relationship between doing work on a body and heating it up, for the brakes that stop a car become hot in the process, and a rapidly moving bullet that is stopped by a target is partially melted.

The exact relationship between work and heat was discovered in the nineteenth century by an English physicist, James Joule. He constructed an apparatus like that shown in

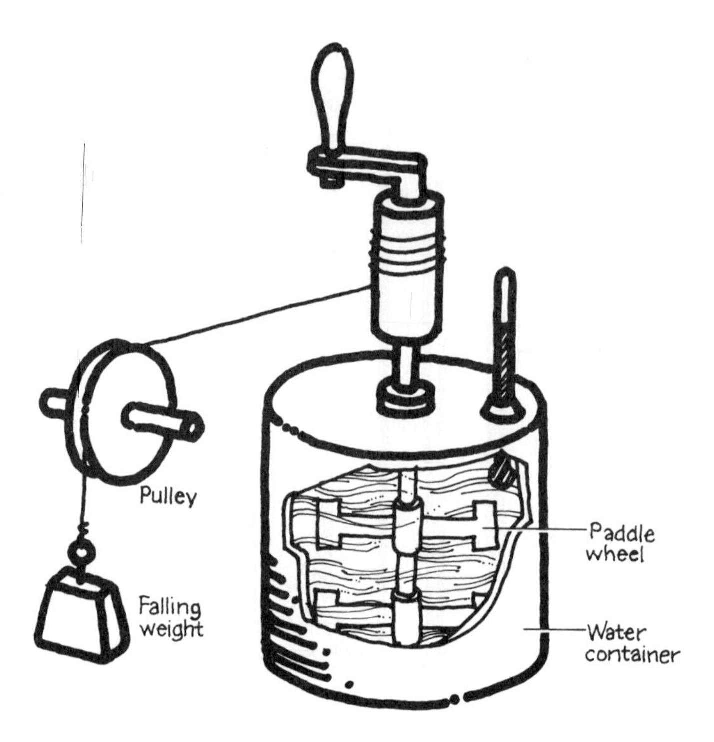

FIG. 10.3

Fig. 10.3, in which a falling weight turned a paddle wheel surrounded by water. Work was done as the weight fell and the wheel turned in the water. Therefore the water was found to get warmer. Joule's measurements showed that a 1-pound weight falling through a distance of 772 feet will raise the temperature of 1 pound of water 1 degree Fahrenheit.

Joule's work clearly demonstrated that heat and energy are equivalent, and from his results we may obtain the "mechanical equivalent of heat." The heat–energy relationship may be formally expressed in the *First Law of Thermodynamics,* which is simply a statement of the law of conservation of energy, with it being understood that heat must be included when one makes any calculations.

The first law of thermodynamics is important because virtually every power source we are familiar with may be treated as a *heat engine.* Broadly speaking, a heat engine is any device that changes heat energy into mechanical energy. Thus gasoline, diesel, and steam engines may all be described as heat engines.

A heat engine may be thought of as operating between two heat reservoirs, one hot and the

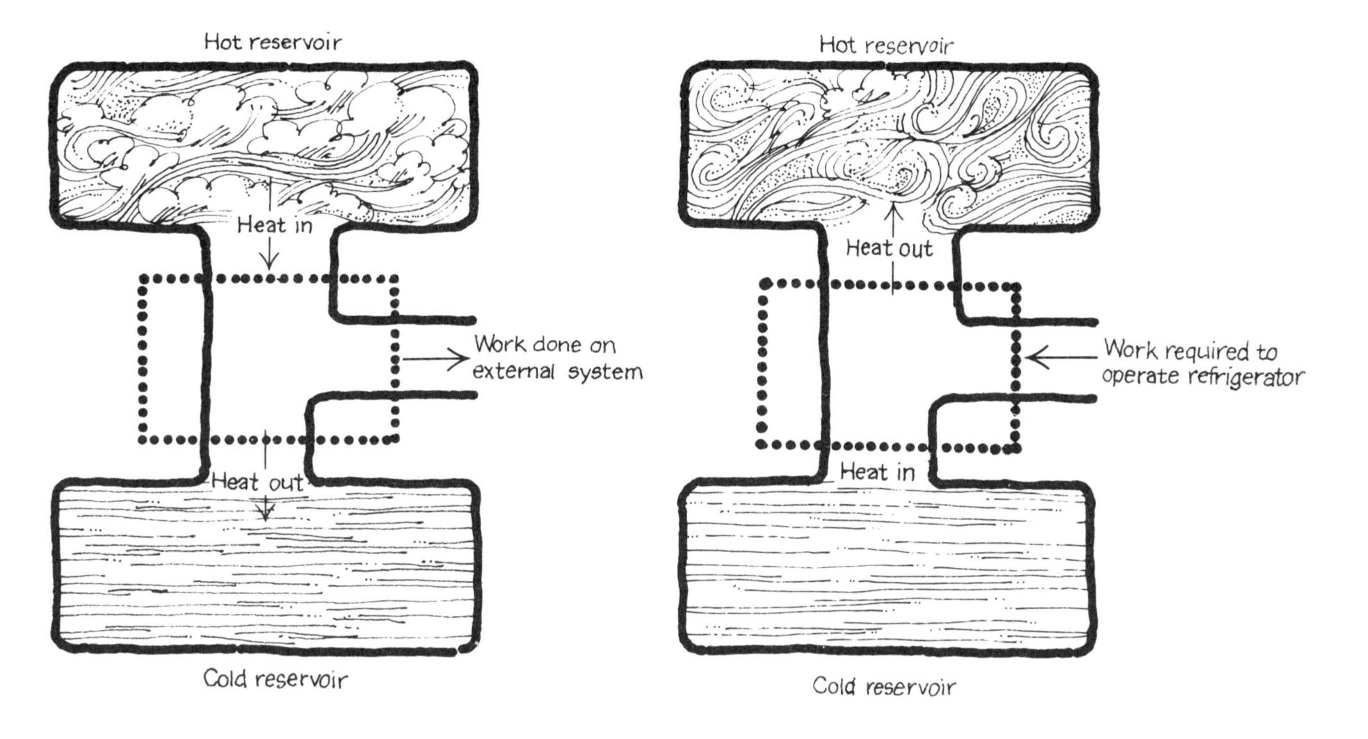

FIG. 10.4 Schematic of the operation of (a) a heat engine, (b) a refrigerator.

other cold. Heat energy is extracted from the hot reservoir, a certain amount of work is done, and the remaining heat energy is discharged to the cold reservoir (Fig. 10.4). The first law of thermodynamics tells us that the total of the *work done* by the engine plus the *heat energy exhausted* must equal the *heat energy taken in*. That is, total energy must be conserved by the system.

But the first law tells us nothing about why an airplane cannot use a heat engine that extracts heat energy from the atmosphere and discharges colder air as the exhaust, or why a ship cannot extract heat energy from the ocean for its power. The reason these things cannot happen is given, however, by a *Second Law of Thermodynamics*. The first law forbids the creation or destruction of energy; the second restricts the ways in which energy may be used. Of course, heat may be extracted from a system and the system left colder than its surroundings. This is what a refrigerator or air conditioner does, but you pay for the work you do in taking heat out of the system. If there were no second law of thermodynamics, you could operate these appliances as sources, not consumers, of energy.

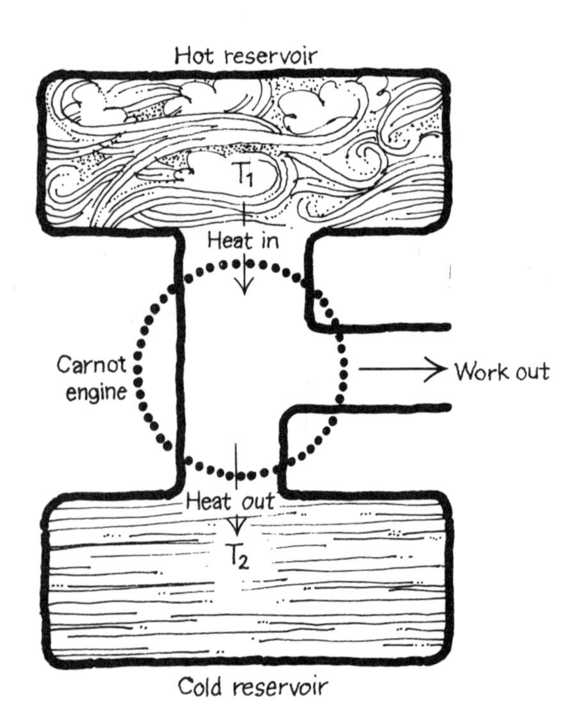

FIG. 10.5 Efficiency of a Carnot engine $= \frac{T_1 - T_2}{T_2}$.

An idealization of the heat engine was described by a Frenchman, Nicolas Leonard Sadi Carnot, during the last century (Fig. 10.5). This engine, operating between two constant-temperature reservoirs, would convert a fraction of the heat energy it extracts from the hot reservoir into mechanical work, and return the remaining heat energy to a cold reservoir. It may be shown that the efficiency of the Carnot engine is equal to the *difference in temperatures between the hot and cold reservoirs divided by the temperature of the cold reservoir.* (Naturally, these temperatures must be expressed in terms of one of the temperature scales for which zero degrees is the absolute zero of temperature.) One can prove that no actual heat engine may operate between two given reservoirs with any greater efficiency than a Carnot engine. Thus the limit of efficiency of a heat engine is determined by the temperatures of the reservoirs available: for a given cold reservoir, the hotter the hot reservoir the greater the efficiency, and, for a given hot reservoir, the colder the cold reservoir the greater the efficiency. In an electric power plant, the temperature of the fuel (coal, oil, gas, or nuclear fuel) determines the temperature of the hot reservoir, while the temperature of the cooler reservoir is determined by its surroundings.

Now it is implicit in the second law that the total heat energy available is not so important as the particular arrangement of that heat energy. For example, if you had two containers of water, one hot and one cold, a heat engine could be operated between them (Fig. 10.6). But if you were to mix the hot and the cold water, even though the same amount of total heat energy would still be present, unless a new cold reservoir existed somewhere, none of that heat energy would be *available* for use by the heat engine.

The arrangement or organization of heat energy is described by the concept of *entropy*. The entropy of a system is simply a measure of the orderliness of that system. If we think of the above containers of water as having the hot—high kinetic energy—molecules separated from the cold—low kinetic energy—molecules, the system is clearly more ordered than if all the molecules were mixed and at the same temperature. Disordering a system increases its entropy.

Now we are prepared to discuss the second law of thermodynamics in terms of its great universal implications: *The entropy of the universe cannot be decreased.* The entropy of some small part of the universe may be decreased (become more ordered) only at the expense of a greater overall increase in the entropy of the universe. The entropy principle, then, suggests a way of defining a universal "arrow of time": one examines two events and determines for which of them the entropy of the universe is greater, and one knows the order in which they occurred.

The illustration of the mixing of two liquids, one initially hot and one initially cold, to form a uniform-temperature mixture is an example of an increase in the entropy of a system. When this entropy was increased, a certain amount of energy was transformed into a form in which it was unavailable. This unavailability of energy

Heat engine could operate between these two "reservoirs."

Total heat energy is the same, but this energy is not available to do work.

FIG. 10.6

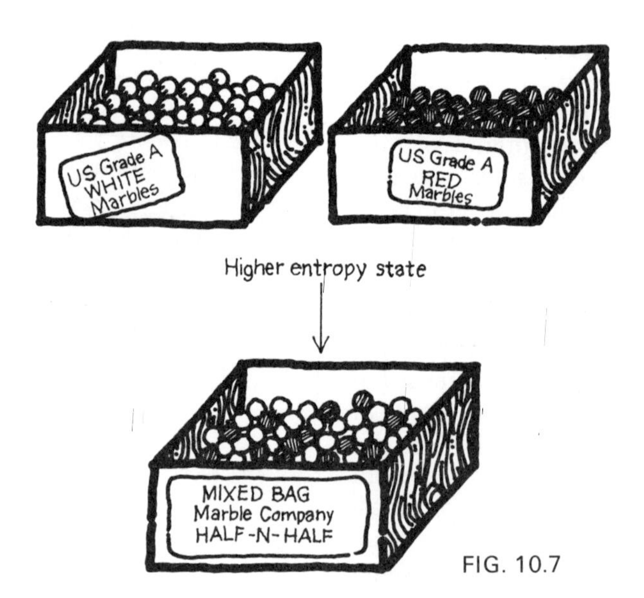

FIG. 10.7

to do work could—in time—lead to the "heat death of the universe": the same amount of energy as always would be present, but none of it would be available for further work.

We can easily think of an increase in the disorder of a system in statistical terms. Imagine that you have two boxes of marbles, one white and one red (Fig. 10.7). One box contains only white marbles, one contains only red. Each box of marbles is plainly ordered and the entropy of our system is at a minimum. Now open the tops, turn the boxes together and shake them. As the shaking proceeds, some of the red marbles bounce into the white box and vice versa. After a while you might expect to find a completely homogeneous distribution of the marbles in which each box would contain half red and half white marbles. But now continue to shake the marbles, stopping occasionally to examine them. When do you expect them to return to their original configuration, with all of the reds in one box and all the whites in the other? Obviously, you don't expect them to "unmix" any more than you would expect to shuffle a deck of cards, already mixed together, and find that they were afterward ordered like a new deck.

Statistically, we may explain these processes by noting that there are many more disordered arrangements of the marbles than ordered configurations. Therefore a disordered state is much more probable than an ordered one. Similarly, there are 8.0658×10^{67} possible arrangements of the 52 cards in a deck, but only one of those is the one found in a fresh deck. (We may even use this principle to explain why a beginning driver may have great difficulty in parallel-parking a car, yet can drive away from that parking space with no trouble: there are many more "unparked" states than parked ones; therefore the probability of a transformation from a parked to an unparked state is much higher than the other way around.)

From our discussion of the Carnot engine, it follows that the only heat engine that is 100% efficient is an engine which has its cold reservoir at the absolute zero of temperature. Now a Third Law of Thermodynamics rears its head. A statement of the third law is that *one can never reach the absolute zero of temperature in a finite number of processes.* Work may all be turned into heat energy, but the reverse process can never quite be achieved. For this reason, the three laws of thermodynamics are sometimes abridged as follows: The first law says you can't win, the second say you can never break even, and the third says you can't even get out of the game.

10.3 THE PRODUCTION OF LOW TEMPERATURES

The production of temperatures lower than those found in nature has proceeded by various routes. Although an apparatus for making ice existed more than 200 years ago, mechanical refrigeration was not invented until 1851, by Dr. John Gorrie, a Florida physician. Dr. Gorrie was motivated to produce this machine to cool the sickrooms of his patients. Techniques for the liquefaction of gases progressed so that by 1898 James Dewar was able to liquefy hydrogen. The last gas to be liquefied was helium. This achievement took place in 1908, and was performed by the Dutch physicist Kammerlingh Onnes.

Mechanical refrigeration is not capable of producing temperatures low enough to liquefy all the gases. There are, however, techniques available to produce temperatures as low as a small fraction of 1 degree. If a compressed gas is allowed to expand freely, it is cooled in the process of expansion. This cooling is known as the *Joule-Kelvin effect.* (It is also sometimes called the Joule-Thomson effect; Lord Kelvin and William Thomson were the same person.) Oxygen and nitrogen are often liquefied by the use of this process. This method of liquefying gases is particularly desirable because the machines involved require no moving parts.

Helium cannot be liquefied by use of the Joule-Kelvin method until it has first been cooled to a temperature below that of liquid nitrogen. Although helium can first be cooled to the temperature of liquid hydrogen and then liquefied by the Joule-Kelvin method, engineers usually avoid using hydrogen because of its flammability. For these reasons the process by which liquid helium is usually produced involves cooling by letting the helium gas do work against a piston. Just as, in a steam engine, the hot vapor (steam) is cooled when it does work (pushing against a piston and making the engine operate), so helium gas is cooled when it does work against a piston which is part of the helium liquefier. When the gas has been sufficiently cooled, it is then liquefied by use of the Joule-Kelvin effect.

After helium has been liquefied (at 4.2 K), its temperature may be reduced still further by evacuating the space above the liquid by means of a vacuum pump. This process of pumping on the helium is effective because the rapid

TABLE 10.1

Event	Temperature, K
Water vapor becomes a liquid	373.2
Triple point of water (defined)	273.16
Carbon dioxide gas becomes a solid (dry ice)	194.7
Oxygen becomes a liquid	90.2
Nitrogen becomes a liquid	77.3
Hydrogen becomes a liquid	20.3
Helium becomes a liquid	4.2
Lambda point of helium	2.2

evaporation of the liquid helium cools it to still lower temperatures. By use of this method, liquid helium temperatures of below 1 K may be reached.

Still lower temperatures may be obtained by a technique known as *adiabatic demagnetization.* A paramagnetic material (one which is not normally magnetic, but which becomes slightly magnetized when placed in an external magnetic field) is magnetized while it is being cooled by liquid helium to the lowest temperature that may be reached in the helium. The paramagnetic material is then thermally isolated from the helium and the external magnetic field is turned off. When this happens, the paramagnetic material is cooled to a very low temperature. Temperatures below 0.001 K have been reached by sequential application of this technique.

In our discussions later in this chapter, we shall be concerned with some of those thermal effects connected with temperatures lower than normal room temperatures. At these low temperatures many interesting and useful phenomena may be observed. To give you an idea of the range of temperatures involved, Table 10.1 lists some important temperatures.

10.4 LIQUID HELIUM PHENOMENA

The third law of thermodynamics tells us that the absolute zero of temperature cannot be reached in a finite number of processes, though temperatures very near the absolute zero can be and have been achieved. In this extremely low temperature region, a number of remarkable phenomena involving liquid helium are observed. We shall describe some of them below.

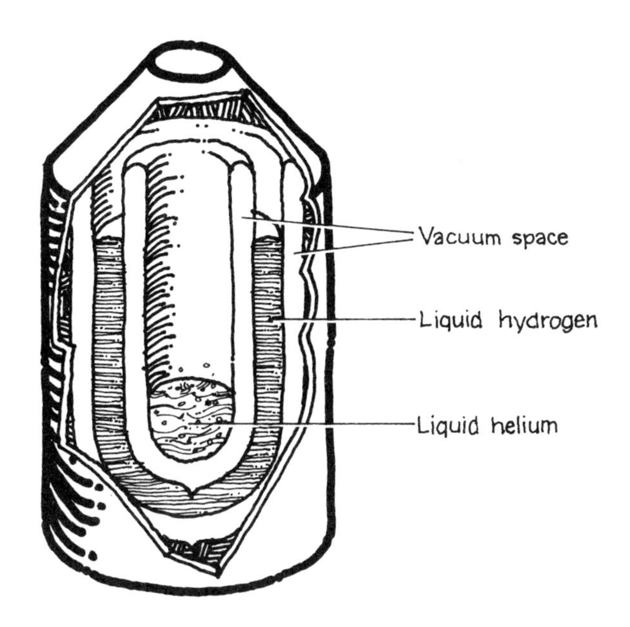

FIG. 10.8 Liquid helium container.

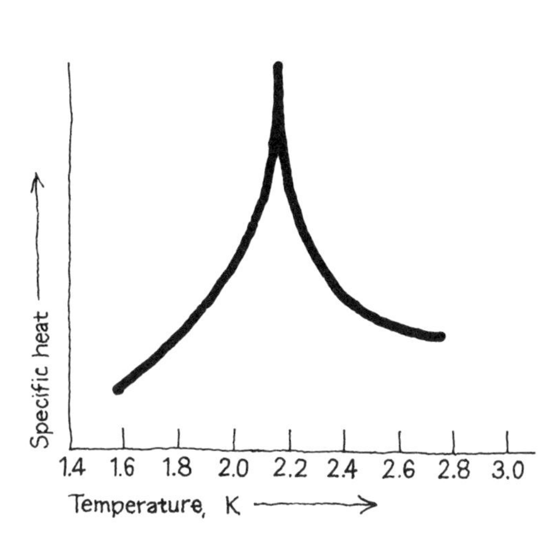

FIG. 10.9 Liquid helium specific heat curve.

After helium becomes a liquid at 4.2 K, it must be protected from gaining heat from outside sources by keeping it in a vacuum-insulated container which is itself usually inside a vacuum-insulated container of liquid nitrogen (Fig. 10.8). Still, heat continually leaks in, and the liquid helium constantly boils away. But as the temperature of the liquid helium is lowered further, a transition takes place at 2.2 K: the boiling motion ceases. Helium is still being vaporized and lost, but the surface appears so calm that it is difficult to see the boundary between the surface of the colorless liquid and the space above it. The point at which this transition takes place is known as the *lambda point* (Fig. 10.9). This name is taken from the Greek letter λ, which looks much like certain curves plotting the change in specific heat of liquid helium near 2.2 K. Below the lambda point, helium occurs in a form that is known as helium II. It is helium II that is involved in the properties that we shall next describe.

Film Creep

Helium II exhibits a property known as *film creep.* A film of liquid helium simply climbs up the inside of the vessel in which it is held and down the outside. This means that the liquid helium levels in two vessels in contact with each other will be equalized, even though the liquid has to flow uphill to bring this about. We pointed out earlier that the temperature of liquid helium cannot be lowered much below 1 K by evacuating the space above it, no matter how rapidly one pumps it. This limitation arises from film creep, for the film of helium that creeps up the sides of the Dewar toward the vacuum

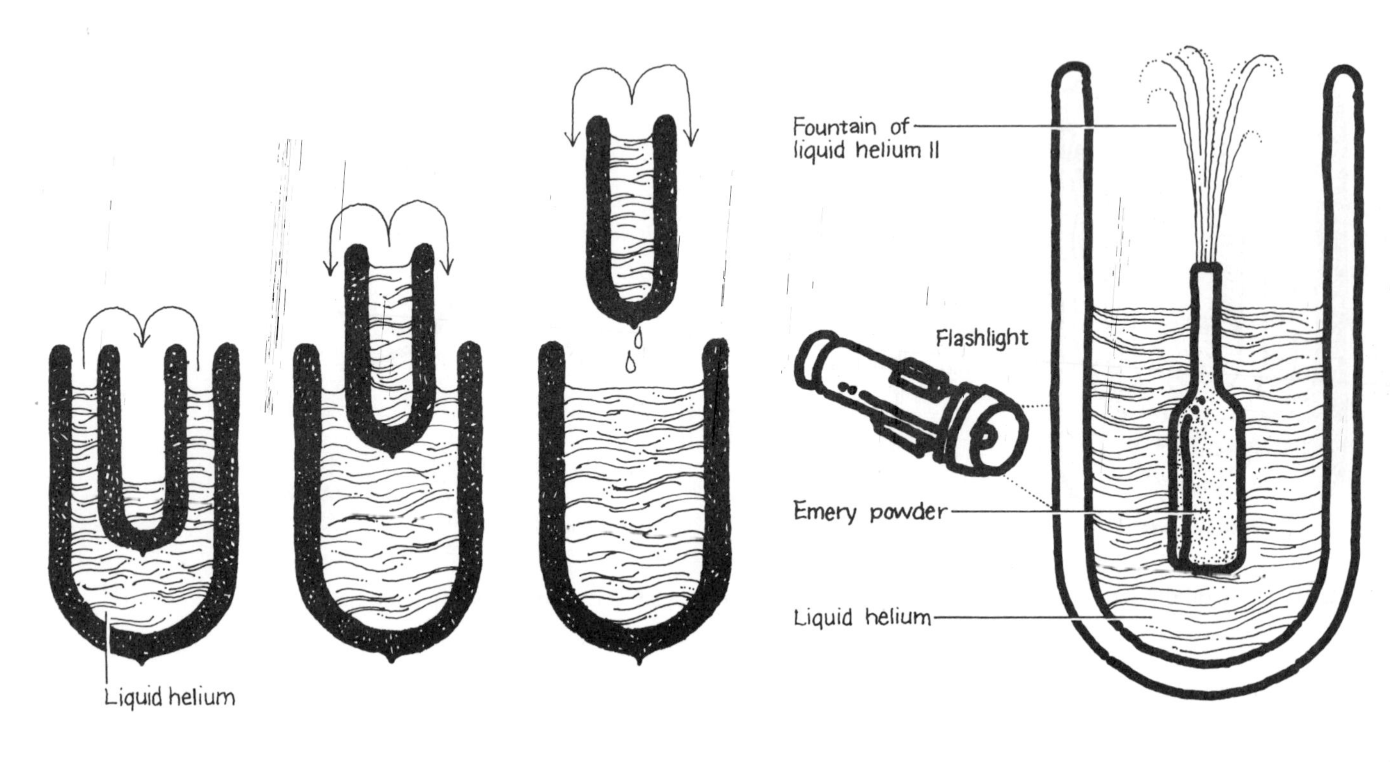

FIG. 10.10 Helium film-creep phenomena.

FIG. 10.11 The fountain effect.

pump constitutes a heat-leakage path into the liquid helium. The illustrations of Fig. 10.10 show some of the ways in which this phenomenon is manifested.

The Fountain Effect

Perhaps the most spectacular of the liquid-helium phenomena is the *fountain effect.* This is observed when a part of a liquid-helium system is closed off from the rest of the system except for a small capillary-tube connector. If the tiniest amount of heat is now added to the liquid helium inside the closed region (one way to do this is to put emery powder in the region and shine a flashlight beam on it), a pressure differential occurs that will result in a fountain of liquid helium. Fountains as high as 30 cm have been observed (Fig. 10.11).

Superfluidity

The above phenomena—film creep and the fountain effect—are actually manifestations of a property known as *superfluidity.* The superfluid form of liquid helium has the unique property of being incapable of exerting shear forces on objects. This means that the viscosity of superfluid liquid helium is zero. Some of the strange effects of zero viscosity that one might observe would include a coin remaining delicately balanced on edge with a high-pressure stream of superfluid liquid helium directed at the coin without knocking it over. (The superfluid liquid helium would simply flow smoothly around the coin.)

A trick question that is sometimes asked is: "Could a fish swim in superfluid liquid helium?" The correct answer is "No," because superfluid liquid helium exists only below 2.2 K and the fish would certainly freeze. But could a "superfish" that would not freeze at that temperature swim in superfluid liquid helium?

10.5 SUPERCONDUCTIVITY

The phenomena described above all involve effects in liquid helium. Now we shall look at some things that happen to other materials that are cooled to temperatures near absolute zero. The first of these phenomena is that of *superconductivity*, which was discovered in 1911 by the Dutch physicist Kammerlingh Onnes. In 1908 Onnes first succeeded in liquefying helium, and by 1911 he was able to use liquid-helium temperatures for other physical research. In that year, while he was making low-temperature measurements of the electrical resistance of very pure samples of mercury, he first observed superconductivity.

We have previously described electrical resistance in materials as being due either to lattice defects—which are temperature-independent phenomena—or thermal effects, which are temperature dependent. If one had an extremely pure sample of a material in which defects did not make an important contribution, as the material

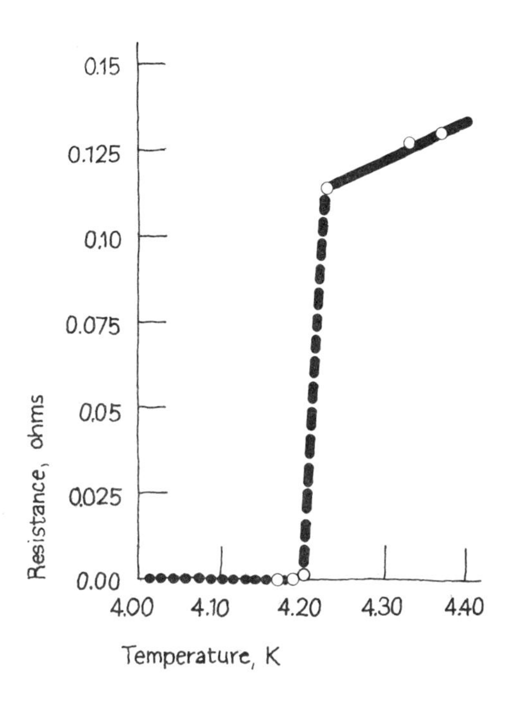

FIG. 10.12 Data on superconductivity, recorded by K. Onnes in 1911.

was cooled toward absolute zero the electrical resistance would be expected to approach zero. This is, no doubt, what Onnes expected to observe in his measurements on mercury.

But to Onnes' surprise, just below 4.2 K, the resistance of the mercury dropped suddenly to a value so near zero that he could not measure it as having any value other than zero (Fig. 10.12). The mercury had undergone a transition to a *superconducting state.* Note that the resistance drop occurred above the absolute zero of temperature, that the transition was sharp and sudden (rather than gradual), and that the resistance drop had been to zero resistance, not to some small value near zero.

There are now 26 elements known to be superconductors, and more than 100 compounds and alloys exhibit this property. The temperature at which presently known superconductors undergo transition ranges from 0.01 K for tungsten to 20.7 K for an alloy of niobium, germanium, and aluminum. Although materials such as tin, lead, and tungsten that are relatively poor conductors (poor, that is, only in comparison with other metals) at room temperatures become superconductors at very low temperatures, silver, copper and gold—all excellent room-temperature conductors—do not. However, aluminum is both an excellent room-temperature conductor and a superconductor at low temperatures.

When we say that the electrical resistance of these materials becomes zero, we do not mean that it reaches some small value near zero; so far as can be measured it actually becomes zero. One experiment designed to see how long a current could flow in a superconducting loop indicated, from the current decay after $2\frac{1}{2}$ years, that it would not be damped out for at least 100,000 years!

There are several important phenomena associated with superconductivity. First, *sufficiently high magnetic fields destroy superconductivity* in a material. This effect limits the amount of current that may be carried by a given superconductor, since the magnetic field set up by the current flowing in the superconductor can be great enough to destroy its superconducting state.

Second, *all magnetic flux is expelled from a superconductor,* a phenomenon known as the

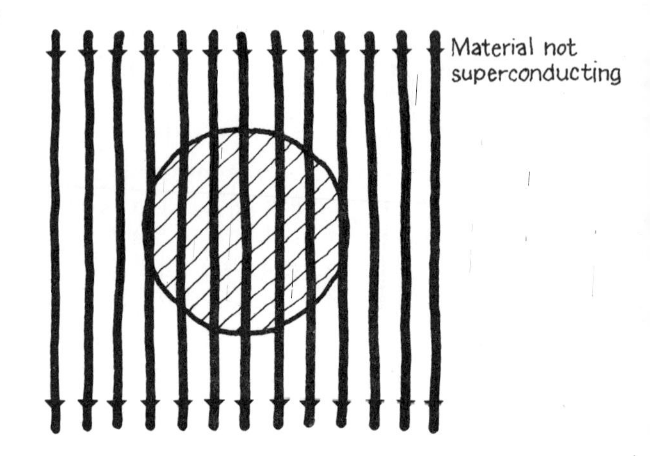

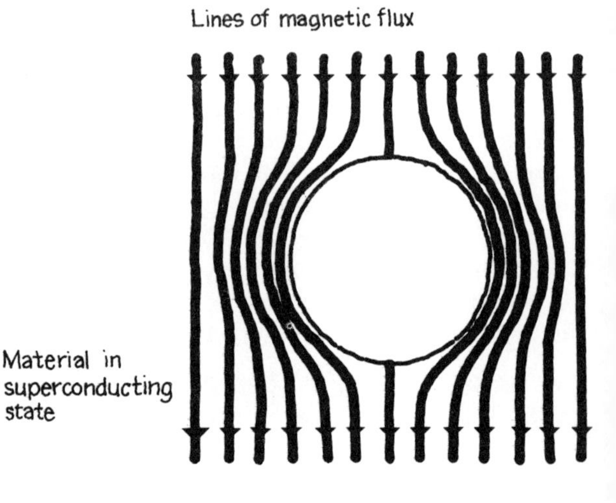

FIG. 10.13 The Meissner effect.

Meissner effect (Fig. 10.13). Because of this, a superconductor can "float" in a magnetic field without mechanical contact with any other surface.

The first explanation of superconductivity that was even partially satisfactory was not advanced until 1957. This was the theory of Bardeen, Cooper, and Schrieffer, known as the *BCS theory*. Modifications to this theory have now produced satisfactory explanations for most of the effects of superconductivity. Because this explanation is highly mathematical, it is difficult to provide a simple qualitative description of it, but it involves the concept that under the conditions associated with superconductivity, electrons may become paired in such a way that they can be propagated through the lattice of the material without experiencing any net impedance. This is possible because, when one of the electrons of a pair has a collision with a lattice site and receives a momentum change in a given direction, the other electron of the pair has a momentum change in the opposite direction, with the result that the net momentum change of the pair is zero.

Even before the theory of superconductivity was understood, many uses had already been made of superconducting phenomena. These uses involved taking advantage of the superconducting characteristics of:

1. Zero electrical resistance.
2. Absence of magnetic flux inside a superconductor.
3. The sharp transition temperature for the onset of superconductivity.

Since a superconductor has zero electrical resistance, large amounts of current may be carried without loss, so long as the magnetic field produced by the current carried is not great enough to destroy the superconducting state. It has been calculated that a single superconducting power line 6 inches in diameter could carry all the power now used by the city of New York. Magnets with far greater field strength than any that can be constructed conventionally are made possible by constructing their coils of superconducting materials. Electrical circuits that need to carry signals without loss can do so using superconducting components, and memory-storage devices may be made that consist of superconducting currents flowing in closed loops which are capable of storing information for centuries.

The fact that there is no magnetic flux inside a superconductor means that magnetic lenses may be made of the flux expelled from a superconductor. These lenses have particular applications for focusing electron microscopes. Frictionless bearings, which are desirable for many applications, are made possible by "floating" a superconducting material in a magnetic field.

Gyroscopes used in inertial guidance systems for space flights and missile operations can now use frictionless superconducting bearings. A superconducting electric motor is a particularly attractive proposition, since it could make use of both frictionless bearings and the zero-resistance properties of the materials from which it was made.

The sharp transition temperature at which a material becomes superconducting makes it possible for very accurate thermal sensors to be constructed using a superconducting material that is normally kept at a temperature just below the temperature at which it makes a transition to the nonsuperconducting state. The incidence of a small amount of heat would raise the

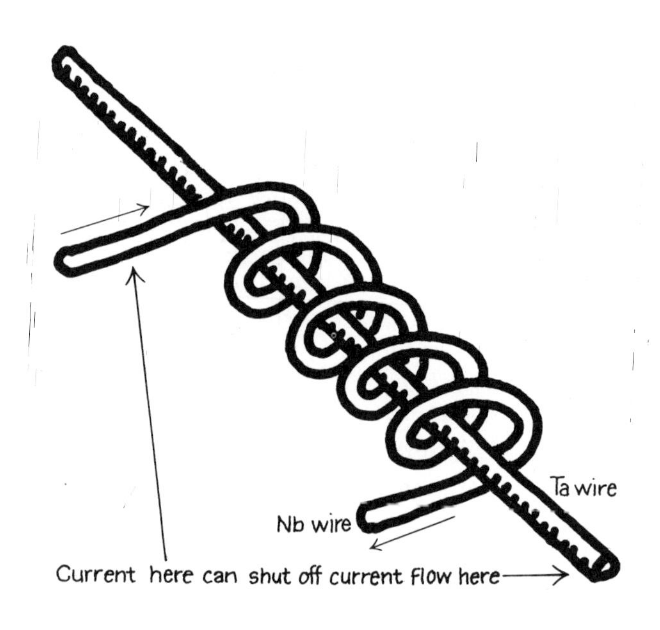

FIG. 10.14 A cryotron.

detector material above its critical temperature, the material would no longer be a superconductor, and the detection of the incident radiant energy would be indicated by an appropriate electrical circuit that would monitor the resistance of the detector material. One possible location for such sensors is in satellites that can detect missile or rocket launchings in any part of the world.

Superconductivity plays an important role in a solid-state electronic device known as a *cryotron* (Fig. 10.14). In this device, a superconductor material such as tantalum (transition temperature = 4.5 K) is surrounded by another material with a higher transition temperature (such as niobium at 9.1 K). A current flowing through the tantalum is cut off by sending a current flowing through the niobium which sets up a magnetic field sufficient to destroy the superconducting state in the tantalum. Thus a cryotron acts like a switch that is capable of switching large currents with minimal power losses.

10.6 OTHER CRYOGENIC APPLICATIONS

Although the phenomena that exist near the absolute zero of temperature include some of the most spectacular known, at higher temperatures (but still well below normal room temperature) there are still many important cryogenic applications.

Most of these applications make use of the fact that a great many natural processes are thermally activated: that is, their rate of activity increases exponentially with temperature. Any such activity may be conveniently represented by an *Arrhenius plot*, in which the logarithm of the given activity rate is plotted against the reciprocal temperature (Fig. 10.15). The straight line indicates that the phenomenon is thermally activated.

For many cryogenic applications a temperature no lower than that of liquid nitrogen (77 K) is required. At this temperature the preservation of otherwise perishable products becomes possible. Whole blood and bone marrow—which is vital for treating persons who have received massive doses of radiation—may be successfully stored for years. The cattle industry in particular has made use of the fact that semen may be readily preserved at liquid-nitrogen temperatures. More than 4,000,000 cattle are

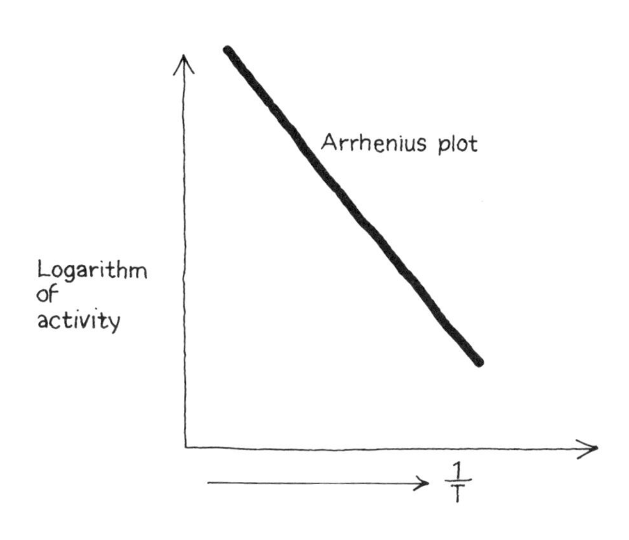

FIG. 10.15 Variation in activity rate for thermally activated phenomena.

born each year in the United States alone as a result of artificial insemination.

In the frozen-food industry there is an increasing use of liquid nitrogen for the quick-freezing of foods. Hot prepared foods that require 3 to 48 hours to freeze by mechanical refrigeration may be frozen in a matter of minutes by spraying them directly with liquid nitrogen. The fact that bacterial activity is stopped almost at once helps these foods remain fresh for much longer periods of time than they would otherwise. Nitrogen is an element particularly well suited for this application because it is odorless and tasteless. Many trucks and freight cars used to transport frozen foods are now cooled by liquid nitrogen rather than by the mechanical refrigeration devices such vehicles previously carried. This use of liquid nitrogen for cooling makes it possible for the frozen foods to remain at lower temperatures while they are being transported than would otherwise be possible.

Medical applications of liquid nitrogen include the development of procedures for the removal of scars and skin blemishes by the application of liquid nitrogen to the affected area. Parkinson's disease, which involves an uncontrollable shaking of the limbs of the body, has been treated by surgically destroying the brain centers responsible for the motion. When performed with a knife this operation can be a rather risky one, for the wrong brain area may be destroyed by mistake. A procedure developed in 1962 involves the use of a probe, only 2 mm in diameter at the tip, which may be cooled by liquid nitrogen. The operation takes place while the patient is conscious and he cooperates in locating the right area to destroy. The cryoprobe is operated in two modes. First, an area of the brain is cooled until it temporarily ceases to function. When the proper brain area has been chosen—this is verified by the patient—it is then further cooled to liquid-nitrogen temperature and thus destroyed. Other successful applications of cryosurgery have involved the destruction of some neoplasms and the removal of organs such as prostates and tonsils in patients with blood diseases.

One of the most dramatic possibilities that has been suggested in connection with cryogenics has to do with the cooling of living persons. If such a technique were perfected, future astronauts might go on long space voyages in a state of "suspended animation" and thereby overcome the time limitations of true deep-space missions. Persons suffering from now-incurable diseases might be frozen and later revived when cures for these diseases had been developed. Or a person who did not like his time and generation might simply wish to finish his life in some future society.

Unfortunately there is no present method of accomplishing anything like a suspended-animation process. Work has been reported on the freezing and attempted restoration of small animals, principally hamsters. For example, in one extensive series of experiments, animals that were frozen for 45 minutes or less, with less than 40% of their total body water being frozen, survived without exception. Animals with 50% of their total body water frozen had a survival rate of about 33% after 10 days, while none for whom more than 55% of the body water was frozen survived. Thus it must be said that, at the present time, even small animals cannot be completely frozen and recovered. Attempts have also been made to freeze whole human body organs, but thus far all of those frozen below −20°C have been nonfunctional shortly after thawing.

The problems that arise in freezing living cells are due to the fact that the water in the cells freezes first. This produces three potential sources of damage to the cells:

1. Those fluids that are not frozen become much more concentrated and the concentrated solutions may damage cell walls.
2. Changing water to ice increases pressure gradients across cell walls, upsetting the osmotic flow.
3. The ice crystals formed may themselves do mechanical damage to cell walls.

In addition to the problems of freezing, the thawing of frozen animals presents added difficulties, since it appears that it is necessary for the whole body to be thawed at approximately the same time. This does not happen in the case of thawing due to the simple application of heat, for the outside of the animal may reach room temperature, while the vital organs remain frozen.

One way of avoiding these problems involves the addition of protective agents. It has been found that certain insects that are particularly immune to cold weather damage secrete glycerol, and lower the freezing temperature of body cells, a sort of built-in antifreeze protection. The recent discovery of what is thought to be an anomalous form of water that does not freeze at 32°F has prompted speculation that it might be substituted for the ordinary water in body cells and thereby permit them to be cooled to much lower temperatures than is now possible. This anomalous water—thought by many to be a polymeric form of water, and thus dubbed *polywater*—boils above 400°F and, rather than freezing, undergoes a change into a gel-like material at −40°F.

Even in the face of the discouraging prospects of recovering frozen bodies, during recent years some persons who have died have had their bodies preserved in liquid nitrogen in the hope that at some future time they may be restored to a living condition. Not only would cures have to be discovered for whatever their cause of death happened to be but the (now) irreversible brain damage suffered after death would have to be repaired. The human brain suffers irreversible damage starting about five minutes after its oxygen supply has been cut off. It is probable that in all current cases of cryogenic body preservation, the allowable time for the brain to be without oxygen has been greatly exceeded.

FOR MORE INFORMATION

Michael McClintock, *Cryogenics,* New York: Reinhold, 1964.

K. Mendelssohn, *The Quest for Absolute Zero,* New York: McGraw-Hill, 1966. (In paperback.)

Harold T. Merryman (editor), *Cryobiology,* New York: Academic Press, 1966.

O. V. Lounasmaa, "New Methods for Approaching Absolute Zero," *Scientific American,* December 1969, page 26.

Theodore A. Buchhold, "Applications of Superconductivity," *Scientific American,* March 1960, page 270.

Gerald Feinberg, "Physics and Life Prolongation," *Physics Today,* November 1966, page 45.

QUESTIONS

1. What is the origin of the Fahrenheit temperature scale?
2. What is the significance of the word "centigrade" in temperature measurements?
3. What has the triple point of water got to do with temperature standards?
4. Is a temperature of 60°F twice as hot as 30°F? Why?
5. Which of the elements becomes a liquid at the lowest temperature?
6. What is the lambda point?
7. Describe film creep, superfluidity, and the fountain effect.
8. To what does superconductivity refer?
9. Does any correlation exist between good room-temperature conductors and superconductors?
10. What effect does a high magnetic field have on a superconductor?
11. Name some superconductor properties that have particularly useful applications.
12. How does a cryotron work?
13. Why is liquid nitrogen especially suitable for the quick freezing of foods?
14. Describe some medical uses of liquid nitrogen.
15. How may body cells be damaged when they are frozen?
16. How successful have the experiments involving freezing and thawing small mammals been?
17. What is the current prognosis for the successful freezing and thawing of humans?
18. How does polywater differ from ordinary water?
19. How is it possible to uniquely specify a temperature simply by "−40"?

PROBLEMS

1. It is both simpler and cheaper to liquefy air (78% nitrogen) than to separate out the nitrogen and liquefy it alone. Since liquid air is only a few degrees warmer than liquid nitrogen, why is liquid air not used widely?

2. Some people are surprised, when viewing a container of liquid nitrogen for the first time, to see that the liquid bubbles violently when a room-temperature object is dropped into it. Further, the vapor visible around the liquid falls instead of rising. Explain.

3. Before World War II the U.S. refused to sell helium to Germany. As a result, the large German dirigibles were filled with the highly flammable gas, hydrogen, and the Hindenberg tragedy followed. Can you think of any similar situation that might result in cryogenic applications?

4. Medical schools have a perennial problem in obtaining enough human bodies for their students to dissect. Since demand now exceeds supply in this market and is likely to get worse, an enterprising group formed Cryo-Cadavers, Inc., to buy corpses, store them in liquid nitrogen and resell them later to medical schools at a profit. Comment on the prospects of the success of this venture.

5. Brass Monkey, Ltd., a firm engaged in the business of storing bodies in liquid nitrogen, found that maintaining the nitrogen levels in their Dewar containers was costing too much. A bright (?) young officer of the company suggested shipping the stiffs to the dark side of the moon, where the temperature is −260°C. What do you think of this idea?

6. If polywater could be substituted for the normal water in human cells, what would then be the prospects of freezing and subsequently thawing out humans?

7. An electrical current that flows in a superconductor will evidently continue to flow indefinitely. From thermodynamics we learn that perpetual-motion machines are not possible. Is a current flowing forever in a superconductor a perpetual-motion machine?

8. Discuss the possible applications and limitations of superconducting power lines that would carry electrical power from distant generating plants to densely populated areas.

9. It is very cold in deep space. Suggest some cryogenic phenomena one might intend to make use of in a space vehicle.

GLOSSARY OF TERMS

Acceleration. The rate of change of velocity, either positive or negative. An object whose velocity increases by 1 foot per second each second is accelerated at 1 foot per second per second. A negative acceleration is often referred to as a *deceleration*.

Adiabatic demagnetization. The removal of an applied magnetic field from a material under conditions that do not allow a transfer of heat. When this effect is used to reach very low temperatures, a paramagnetic material is cooled to the lowest possible temperature with a strong magnetic field applied. When the paramagnetic material has been both cooled and magnetized to the greatest degree possible, the space around the material is evacuated—to provide insulation against the transfer of heat—and the magnetic field is removed. The paramagnetic material's temperature then falls to a lower value than before.

Absolute zero of temperature. The lowest temperature that exists, although the Third Law of Thermodynamics tells us that this temperature cannot be reached in a finite number of processes. The statement is sometimes made that all molecular motion ceases at absolute zero, but this is not correct. Rather, molecules possess their very lowest, or *zero-point* energies, at the absolute zero of temperature.

Alpha particle (α*-particle*). A subatomic structure consisting of two protons and two neutrons that may be emitted by nuclei undergoing radioactive decay. This extremely stable structure of two protons and two neutrons forms the nucleus of the helium atom.

Angstrom unit (Å). A measurement of length, equal to 10^{-8} centimeter (about 4 billionths of an inch). The wavelengths of the visible portion of the electromagnetic spectrum lie between 4000 Å (violet) and 7000 Å (red).

Aphelion. The point farthest from the sun on the elliptical path of a body in orbit about the sun.

Apogee. The point farthest from the earth on the elliptical path of a body in orbit about the earth.

Astronomical unit (AU). The mean distance between the earth and the sun: 92,956,000 miles.

Atom. The smallest structure in which an element retains its physical and chemical identity. The atom consists of a very small, positively charged nucleus about which electrons orbit. The number of protons in the nucleus of an atom is equal to its *atomic number*, while the number of its protons plus the number of its neutrons is equal to its *atomic weight*. The elements found in nature have atomic numbers 1 through 92. A neutral atom is one whose number of electrons equals its atomic number.

Atomic number. The number of protons in the nucleus of an atom.

Atomic weight. The total number of protons plus neutrons in the nucleus of an atom. The atomic weight of an element is written either as a superscript following the symbol for the element (U^{235}) or as a number following the symbol for the element (U-235).

Baryon. A family of elementary particles whose *number* always remains constant (law of conservation of baryons). The baryons include the nucleons (protons and neutrons) and the hyperons. Baryons are a part of the hadron family.

Balmer lines. A series of spectral lines of hydrogen that includes lines visible to the eye. These lines represent transitions between the first excited energy level of the hydrogen atom and its higher energy levels.

Beta decay (*β-decay*). A radioactive decay of a nucleus that involves emission of a beta particle (electron or positron). Beta decay involves the "weak" force. The nucleus that remains after beta-minus (electron) emission has one more proton and one less neutron than before. The nucleus that remains after beta-plus (positron) emission has one less proton and one more neutron than before.

Beta particle (*β-particle*). An electron or positron that is emitted from the nucleus of an atom undergoing radioactive decay.

Binary system. The number system that uses the base 2. Since it has only two characters (0 and 1), the binary system is particularly adaptable for use with electronic circuits, in which one digit may be represented by "off" and the other by "on."

Blackbody radiation. The electromagnetic radiation emitted by an ideal radiator, known as a *blackbody.* The intensity of the radiation from a blackbody is characterized by its temperature and wavelength and is independent of the size or shape of the radiator.

Calorie. The quantity of heat needed to raise the temperature of 1 gram of water from 14.5° C to 15.5° C.

Carnot engine. An idealized heat engine that would operate with the maximum possible efficiency between any two thermal reservoirs. Although no Carnot engine exists, it is a useful conceptual device to which one may compare the efficiencies of real heat engines.

Celsius scale. The temperature scale having fixed points at the ice point (0° C) and the steam point of water (100° C), with 100 equal degree intervals in between. This name is now applied to the temperature scale commonly known as the *centigrade scale.*

Centigrade scale. (*See* Celsius scale.)

Centimeter. One hundredth the length of a meter, or about 39/100 of an inch.

Conservation of momentum. The physical principle that, in the absence of any net external force, the total momentum of any system does not change. Conservation of momentum is one of the most fundamental laws of physics.

Cosmic rays. Highly energetic waves and particles that reach the earth's atmosphere from outer space.

Deuterium. An isotope of hydrogen having one proton and one neutron in the nucleus.

Deuteron. A subatomic particle made up of one proton and one neutron.

Diffraction. The result of the interference of one part of a wave with another part of that wave when the wave passes through a narrow opening. The pattern produced consists of alternating areas in which the parts of the wave add together in phase and reinforce one another and areas in which the waves add together out of phase and are attenuated. A diffraction pattern of light appears as alternate bright and dark zones.

Eccentricity. The ratio of the distance of any point on an ellipse from the focus to the distance of that point from a fixed line. This ratio is always less than 1. The greater the value of the eccentricity, the more "elliptical" the ellipse appears.

Ecliptic. The plane of the earth's orbit about the sun.

Electron. An elementary particle, the negative charge carrier that orbits the nucleus in an atom.

Electron volt. The energy gained by an object, with a charge equal to that of the electron, which is allowed to accelerate through a potential difference of 1 volt. The energies of elementary particles are often expressed in terms of electron volts (eV), thousands of electron volts (keV), millions of electron volts (MeV), or thousands of millions of electron volts (GeV).

Electromagnetic spectrum. The entire range of electromagnetic waves, extending from the longest low-frequency waves (with wavelengths measured in miles) beyond the shortest gamma rays (with wavelengths of less than one-billionth the wavelength of visible light). The electromagnetic spectrum also includes thermal radiation, radio waves, the infrared, the visible, the ultraviolet, and x-rays.

Electromagnetic waves. Waves that are characterized by both an electric and a magnetic field. These waves include—but are not limited to—the visible light rays. The product of the frequency and the wavelength of a wave is equal to the speed with which it travels. Electromagnetic waves travel at *c*, the speed of light in free space.

Ellipse. The locus of a point, the sum of whose distances from two fixed points is a constant. An ellipse is generated by slicing through a cone at an angle less than that of the side of the ellipse. A circle is a special form of an ellipse.

Entropy. A thermodynamic function that may be used to describe the state of a system. The Second Law of Thermodynamics states that in any nonreversible process the entropy of the universe always increases. Entropy is a measure of the randomness, or disorder, of a system.

Erg. A very small unit of work, equal to 2.778×10^{-14} kilowatt-hours.

Equipartition of energy. The distribution of the energy of a system in equal amounts to each of the active degrees of freedom of the system. For example, a single atom has three degrees of freedom (it may move in three directions that are perpendicular to one another). A pair of atoms (a diatomic molecule) has five degrees of freedom (in addition to their translational motion in three directions, the two atoms may *rotate* about one another and *vibrate* toward and away from one another).

Fahrenheit scale. The temperature scale based on fixed points at the ice point (32° F) and the steam point (212° F) of water, with 180 equal degrees in between.

Gamma-ray (*γ-ray*). A photon emitted by a nucleus when it decays from an excited state.

Geodesic. The shortest line between two points on a surface. On a plane the geodesic is a straight line. but on a sphere it is the arc of a circle—including the two points—whose center is at the center of the sphere.

Gravity. One of the four basic forces in nature. Gravity is one aspect of the attraction which one mass has for another. The gravitational force is the weakest of the forces of nature.

Hadron. The family of elementary particles that interacts by means of the strong (nuclear) force. The hadrons include both baryons and mesons.

Half-life. The time required for half of a given sample of radioactive material to decay (its level of radioactivity decreases by one-half). Regardless of the age of a sample of radioactive material, its rate of radioactivity decreases by 50% during a time period equal to the half-life of the material.

Helium. The element whose nucleus contains two protons. The most abundant isotope of helium contains two protons and two neutrons.

Hertz. The term now used for "cycles per second." A frequency of 60 cycles per second (60 cps) is now written 60 hertz (60 hz). "Hertz per second" is *not* a unit of frequency.

Hyperon. The family of baryons with masses greater than the masses of the nucleons. Hyperons are characterized by a *strangeness number*, while the nucleons are not.

Hydrogen. The element whose nucleus contains one proton. Hydrogen has an isotope with one neutron (*deuterium*) and an isotope with two neutrons (*tritium*).

Inertial system. A reference system in which a body at rest, not acted on by any net force, remains at rest. Any nonaccelerated system of reference may form an inertial system, as, for example, a vehicle in deep space, far from any massive body, drifting with its motors cut off. The spinning earth is not actually an inertial system, although its rotation is so small that, for many purposes, it may be considered as an inertial system.

Infrared. That part of the electromagnetic spectrum that extends from the long-wavelength limit of the visible spectrum (the red) through wavelengths longer than those that may be detected by the human eye.

Ion. An atom which has unequal numbers of protons and electrons, and which hence carries a net electric charge. The removal of one or more electrons from a neutral atom results in the formation of a positive ion, while the addition of one or more electrons to a neutral atom creates a negative ion.

Interference. At any point at which two or more waves come together they may be said to interfere. The result of this combination of waves may be *constructive* (they add together to produce a larger displacement, the sum of the individual displacements) or *destructive* (they may completely cancel each other). Light from an appropriate source falling on a series of narrow slits emerges with the rays coming from each slit interfering with the rays from each other slit. The result is a pattern of alternately bright (constructive interference) and dark (destructive interference) areas.

Isotope. Nuclei having the same numbers of protons but different numbers of neutrons are all *isotopes* of the element characterized by their common proton number.

Kaon. The k-meson, an elementary particle with a mass 966 times that of the electron. Kaons are found in nature with either positive or zero electric charge.

Kelvin. The unit of temperature, defined as 1/273.16 of the temperature of the triple point of water. The kelvin replaces "degrees Kelvin"; thus the lambda point of liquid helium is specified simply as "2.2 kelvins," rather than "2.2 degrees Kelvin."

Kinetic energy. The product of one-half the mass of an object and the square of its velocity. To stop a moving object, an amount of work must be done on the object equal to its kinetic energy. For example, when a moving automobile hits a stone wall, its kinetic energy is equal to the work done in stopping the car, primarily in the form of bending body panels (and probably occupants). In happier circumstances the kinetic energy of the moving auto is dissipated by the work done by the braking system.

Light year. The distance light travels through free space in a year: 5,878,500,000,000 miles.

Lepton. The family of particles that interact by means of the weak interaction. The leptons include the electron, the electron neutrino, the muon, and the muon neutrino.

Lyman lines. The series of hydrogen spectral lines that involves transitions between the ground state and the excited energy levels of the hydrogen atom. These lines are in the ultraviolet part of the spectrum.

Mass. A measure of the quantity of matter in a body; the ratio of the force exerted on a body to its rate of change of velocity.

Meter. The fundamental unit of length in the metric system of measurement. The meter, once defined as the distance between two scratches on a platinum-iridium bar, is now defined as 1,650,763.73 wavelengths of the orange-red spectral line emitted by krypton-86.

Momentum. The product of a body's mass and velocity. In the absence of a net external force acting on any system, its momentum remains constant.

Muon. The mu-meson (μ-meson). A lepton with 207 times the electron's mass but having the same electrical charge.

Neutrino. A lepton with a mass approaching zero. There are two types of neutrinos: those associated with electrons and those associated with muons.

Neutron. A nucleon having zero electric charge. The neutron is about $1\frac{1}{2}$ electron masses heavier than the proton. Neutrons interact by the strong force.

Nucleon. A proton *or* a neutron in the nucleus of an atom. These particles are nearly identical in mass, and their behavior in the nucleus, under the influence of the charge-independent nuclear force, is similar in many respects.

Nuclear force. The strongest of the forces in nature. The nuclear force is responsible for binding the nucleons together to form the atomic nucleus. This force acts independently of electrical charge and is very short-ranged.

Overtone. A frequency that is twice, three times, four times, . . . , the lowest frequency (or *fundamental*) produced by a vibrating material. The *first overtone*, which has twice the frequency of the fundamental, is also known as the *second harmonic* (the fundamental is the first harmonic). The *second overtone*, with three times the frequency of the fundamental, is identical to the *third harmonic*, etc.

Parabola. The locus of a point that is equidistant from a fixed point and a fixed line. A parabola may be generated by slicing through a cone parallel to one side of the cone.

Pair production. The process by which a sufficiently energetic photon, near a nucleus, may be transformed into an electron and a positron. This process involves the transformation of energy (the photon, a quantum of electromagnetic radiation) into matter (the electron and positron).

Paramagnetic. A material that is not normally magnetic but may be made to act like a magnet in the presence of an applied magnetic field.

Parity. A means of describing symmetry with respect to inversion in space. An inversion in space is accomplished by replacing all spatial coordinates by their negatives, which is the same thing as reflecting spatial coordinates through the origin. The *conservation of parity* implies that there is no way to tell whether one is observing a real physical experiment or its mirror image. Parity is not conserved under the weak interaction.

Perigee. The point nearest the earth on the elliptical path of a body in orbit about the earth.

Perihelion. The point nearest the sun on the elliptical path of a body in orbit about the sun.

Photoelectric effect. The emission of electrons from a material when light of sufficiently high frequency falls on it. In this process the energy of the photon is transferred to the electron, giving it enough energy to escape from the material in which it was contained.

Photon. The quantum of electromagnetic radiation. Photons travel at the speed of light and have no rest mass.

Pion. The pi-meson (π-meson). An elementary particle having a mass 273 times that of the electron. Pions are found in nature either positively or neutrally charged. The neutral pion (pi-naught) is its own antiparticle.

Planck's constant. The ratio of the energy of a photon to its frequency. This is one of the fundamental constants in nature. Planck's constant is denoted by the letter h or, when divided by 2π, by $\hbar$.

Plutonium. A synthetic element with 94 protons in the nucleus. Pu^{239} releases a great deal of energy when it undergoes nuclear fission.

Positron. The antiparticle of the electron. The positron carries a positive electrical charge that is precisely equal in magnitude to the charge of the electron.

Proton. A fundamental particle found in the nuclei of atoms. The proton has 1836 times the mass of the electron and it carries a positive charge equal in magnitude to the charge of the electron. Protons, along with neutrons, in nuclei are referred to as *nucleons*. Protons interact by means of both the nuclear force and the electromagnetic force.

Quark. A postulated family of three particles that would carry charges whose magnitudes would be $\frac{1}{3}$ and $\frac{2}{3}$ that of the charge of the electron. If quarks exist, protons and electrons and other particles could be constructed of quarks.

Quasar. Short for "quasi-stellar object." Quasar spectra show extremely large red shifts, and quasars are therefore thought to be very distant objects. If they are as distant as they appear, quasars emit enormous amounts of energy, and their energy production cannot be accounted for by any known mechanism.

Rad. A radiation unit equal to an absorbed dose of any ionizing radiation of 100 ergs per gram of absorber.

Rem. A radiation dose equal to the dose in rads times the Relative Biological Effectiveness (RBE) factor of the radiation. The origin of the name is "roentgen equivalent man."

Roentgen. A radiation unit defined so that this amount of radiation in 1 cubic centimeter of dry air at 0° C and 760 millimeters of mercury pressure produces 2.083×10^9 ion pairs.

Seismometer. An instrument that responds to movements in the medium to which it is attached. On earth seismometers measure the intensity, direction, and duration of earthquakes.

Spectrometer. An instrument used to analyze electromagnetic radiation by wavelength. Radiation of many different wavelengths entering a spectrometer is spread out in such a way that each wavelength appears at a separate location. This spreading-out of the electromagnetic radiation may be done by a prism or by a finely ruled grating.

Speed of light. The universal speed limit: 186,282.1 miles per second in free space. A universal constant, *c*. No object with a finite rest mass may ever travel at the speed of light.

Superconductivity. The state in which certain materials conduct electricity with zero electrical resistance at temperatures near absolute zero.

Triple point of water. That point at which ice, water, and water vapor all coexist in equilibrium. This point is defined as 273.16 K.

Tritium. The isotope of hydrogen which has 2 neutrons in the nucleus.

Ultraviolet. That part of the electromagnetic spectrum with wavelengths just shorter than the shortest wavelengths to which the human eye is sensitive (4000 Å).

Velocity. The rate of change of displacement. The speed of an object in a given direction.

Weak force. The third-strongest of the four fundamental forces of nature. The weak force is associated with beta decay and the nonconservation of parity.

INDEX

GREEK ALPHABET

Α	α	Alpha
Β	β	Beta
Γ	γ	Gamma
Δ	δ	Delta
Ε	ϵ	Epsilon
Ζ	ζ	Zeta
Η	η	Eta
Θ	θ, ϑ	Theta
Ι	ι	Iota
Κ	κ	Kappa
Λ	λ	Lambda
Μ	μ	Mu
Ν	ν	Nu
Ξ	ξ	Xi
Ο	o	Omicron
Π	π	Pi
Ρ	ρ	Rho
Σ	σ	Sigma
Τ	τ	Tau
Υ	υ	Upsilon
Φ	ϕ, φ	Phi
Χ	χ	Chi
Ψ	ψ	Psi
Ω	ω	Omega

BCDE7987654321